THE PLENITUDE

SIMPLICITY: DESIGN, TECHNOLOGY, BUSINESS, LIFE

John Maeda, Editor

The Laws of Simplicity, John Maeda, 2006

The Plenitude: Creativity, Innovation, and Making Stuff, Rich Gold, 2007

THE PLENITUDE

Creativity, Innovation, and Making Stuff

RICH GOLD

The MIT Press
Cambridge, Massachusetts
London, England

For information on quantity discounts, email special_sales@mitpress.mit.edu.

This book was set in Scala and Scala Sans by The MIT Press.
Printed and bound in the United States of America.

Library of Congress Cataloging-in-Publication Data

Gold, Rich.
The plenitude : creativity, innovation, and making stuff / by Rich Gold ; foreword by John Maeda.
    p.   cm.
Includes bibliographical references.
ISBN 978-0-262-07289-2 (hardcover : alk. paper)
1. Creative ability.   2. Creative thinking.   3. Technological innovations.
4. Material culture.   5. Materialism.   6. Consumption (Economics)   I. Title.
BF408G563   2006   153.3'5—dc22   2007002120

10 9 8 7 6 5 4 3 2 1

Dedicated to Marina LaPalma and Henry Goldstein

CONTENTS

CONTENTS

I had the fortune of meeting Rich Gold at a National Academies of Science meeting held in San Francisco in 2002. Rich was the invited dinner speaker for our small study group, which was engaged in looking at the intersection of information technology and creativity. It was there that I first encountered "the Plenitude" and the philosophy of Rich's "four hats of creativity": scientist, artist, designer, and engineer. My immediate impression on hearing his lecture was one of delight as I realized I was staring at a real anomaly in the world—a person who had crossed many worlds and lived to tell the tale of his wearing of the four hats. No matter how he leaned while telling his stories, none of the hats ever fell off of his head. It was as if each hat were specially fitted to each of his four personalities.

A few months later, I heard that Rich had passed away. This came as a shock to me, not because I knew him well personally, as I regretfully didn't, but because the world had lost such a rare resource.

Rich was a pure hybridized thinker in a world where monospecialties are the norm.

Achieving simplicity in design is a complex journey that requires many skills spanning from the wildly creative to the purely practical. Good design is something that is best defined when not prescribed, but asked as a series of questions—as Rich asks aloud in this book. Rich's lifework is his unique understanding of the world as something so simple as making more stuff, but also asking the complex questions of why we still bother to make stuff at all. The answers are still to come from all of us. Get ready for the open slate of the Plenitude that awaits you.

PREFACE

This book was written during the last year of Rich Gold's life. The illustrations are selected from those he prepared for the many lectures he gave to a wide diversity of audiences over the past decade. Drawing on his experiences, he summarizes in a playful yet profound way his ideas about art, science, design, and engineering and how these produce "the stuff of the junk tribe" or global corporate consumer capitalism. *The Plenitude* is a graphic textbook, a cartoon treatise, a speculative autobiography. It is also a very practical essay in moral philosophy, rich with ideas and feeling. It is my hope that publication of this work by MIT Press will make this richness (pun intended) known to a wider audience.

Given the volume of Rich's output, the stretch of his imagination, and the diverse arenas in which he pursued his activities, I'd like to give the reader some background and context for pondering the questions that he touches upon.

THE AVANT GARDE TRADITION

Rich's highly original work is deeply rooted in the tradition of the avant garde: the leading edge of a group of practices developed in early twentieth-century Europe under the rubric of modernism. An interplay between originality and continuity is the key to understanding his long strange trajectory: dadaist performer, faux scientist, computer geek, toy designer, cartoonist, project manager, museum exhibit producer, corporate branding theorist, futurologist, and World Economic Forum Fellow. Rich embodied universal roles such as artist-as-trickster, which is present in avant garde art but harks back to ancient sources. Someone with access to the more ineffable experiential realms who also has the gift of sharing those visions with others, whether through story, song, dance, architecture, sculpture, mathematical equation, or any combination of these, merges the functions of artist, scholar, and priest.

Gordon McKenzie, in his book *Orbiting the Giant Hairball: A Corporate Fool's Guide to Surviving with Grace,* talks about his long career at the enormously successful Hallmark greeting card company, where he acts as a sort of shaman, pushing the edges of what people can conceive, challenging the locked-in practices, assumptions, and restraints (the *hairball*) to release creative energy. Seen as everything from a spiritual advisor to saboteur of meetings or even charlatan, McKenzie feels that he provides creative dissonance, serving as an irritant or disruptor to the business-as-usual approach that in any large organization works against innovation. When Rich visited Hallmark, McKenzie recognized Rich as also being an *orbiter,* someone with the ability to work within the constraints of a corporate culture while connecting it to larger cultural patterns, new ideas, different ways of conceiving the business at hand and the business of the future.

Having made more than a few people enormously wealthy, thinking outside the box became the self-important mantra of pre- and postmillennial silicon valley. But most often the person who dares to do it is slapped down by risk-averse management, quashed by economic realities, or rendered inert by a climate of *deja-cool*. As MacKenzie put it, an organization can "officially laud the generation of new ideas while covertly subverting their implementation." Like some dadaist provocateur, Rich was frequently able to slightly stretch the membranes of the box and nudge people into questioning what the box is and what it holds.

A PRODUCT OF HIS TIME

Rich was born in 1950 in suburban America, popularly understood in the United States as a time and place of optimism and prosperity, good free public education, eradicated diseases, and happy childhoods. But as many scholars of popular cultural history tell us, the 1950s were also a time of deep anxiety. Nuclear threat and the Communist menace were an undercurrent of the carefree affect of that decade. In the following years a growing domestic civil rights movement and an unpopular foreign military intervention in southeast Asia contributed to the national disequilibrium. Conditions at the time one "comes of age" always leave an indelible mark.

"Rich Gold" came into being in Buffalo, New York, in 1963 when thirteen-year-old Richard Goldstein, deciding to become a writer, clipped a few letters off the ends of his name to give it more zip. (Not wanting to waste the left-over letters, he created an anagrammatic alter ego, Ned Sarti, who later published numerous works of pulp fiction. But that is another story.) By the 1970s, Rich had found his niche in the musical sector of the avant-garde art world, where concept and execution were paramount; materials, genre, and context could be drawn from the most diverse and disparate sources and combined in any way.

Rich and I met around 1977 at Mills College, where I was getting a degree in poetry and recording media. Mills, a women's undergraduate college, admits men in some of the graduate programs and he was in the graduate electronic music program at the Center for Contemporary Music. This was a time of maximum cross-fertilization among genres, media, and disciplines. Here is an anecdote that concisely illustrates this multimedia interdisciplinary ethos: Rich's 1978 MFA thesis *Concert* was a display of several hundred very small watercolors illustrating a mathematical series and its inversions; these were painted on the backs of several decks of regular playing cards. There was also a video-feed showing Rich performing, with Ed Holmes, inside the empty concert hall. In 1983, my thesis *Exhibit* for an MFA in visual art from U.C. San Diego consisted of an evening performance in concert of my songs in invented and fractured languages and electronic processed vocal pieces incorporating various types of scales and world music modes.

Of course the tendency of cultural trends to tighten up, shrink back, and close off boundaries is always at work too, so this openness couldn't last. Based on our experiences in the art scene of the late 1970s and early 1980s, Rich and I developed an analysis of the social forces and technological givens operating in the production of art at that time. Our argument went against the grain to the highly touted total freedom of the artist. "Clay from the Riverbank: Notes on Curating Computer Art in the '80's" appeared in *Ear* magazine in 1984 and triggered many heated late-night discussions.

THE TWO CULTURES

The intellectual divide (articulated in by C. P. Snow's 1963 book) between the cultures of science and the humanities simply did not exist for Rich. He straddled the worlds of math and language and was one of the earliest adopters of the microprocessor as a tool for art-making, and ultimately for *thinking*. The teenage novelist and young composer Rich Gold was interested in structures and processes, patterns, the interplay between rules and chance, game theory, complete sets, symbol systems, permutations, variations or transformations, series generated by the recombination of a set of elements. His works might be sonic or graphic in nature, incorporating linguistic symbols or "found" artifacts such as thrift store toys and game pieces. "Newton's Nightmare or Como Unbound" (with Paul Robinson) was an installation at San Francisco's Langton Arts. A Rube Goldberg–like assemblage was held together by strings, which, in performance, were individually cut to release all kinds of planned kinetic, electric, sonic, and visual mayhem. "Return to Common Ground" (with Paul Wilson), was an environmental performance for sailboat, fishing poles, mime, and megaphones at Fort Mason in San Francisco Bay.

In the early 1970s Rich pursued "research" as art practice. Using a variety of graphic languages, he created cultures with myths and maths and languages of their own; and, with the Kim1's 1K of memory, allowed them to learn and adapt over time. Every work produced was cogenerated or coenacted with the others. Each project fed into the next, linked as progenitor or offspring, core or elaboration, reversal, inversion, or reiteration in different media: The musical score is generated from a variant of the Fibonacci number series; the title of the song is later incorporated into a live writing event using a tiny computer, which produces a book in a day, written on butcher paper and posted on the walls; the text is later given in performance with singer, balalaika, and electronic space drums in a punk club.

Drawings are turned into symbols for a math equation that generates a pattern that is allowed to expand and form a story. Portraits of the story's characters are painted on panels and shown at Sushi art gallery in San Diego. A *Scientific American* article discussed Rich's algorithm to generate sounds and send them to the other computers of the legendary mid-1970s microprocessor quartet, the League of Automatic Music Composers. This algorithm was also used to choreograph the "social dance" of fifteen artists holding clusters of colored balloons on the hillside vineyard of an art collector in northern California one afternoon in 1978.

Each piece was explained, described, or re-enacted as a Goldograph, a series of pamphlets issued to subscribers over the years. One early Goldograph, *I Primi Lavori di Rich Gold*, was a catalog of works to date, reminiscent of Marcel Duchamp's piece "Valise," containing small replicas of his own previous artworks. "The Cybernetic Oracle," a ten-parameter linear control model that works only when not believed in, was used to tell fortunes in performance and became another Goldograph. "Georgeo of the Suburbs" was a particularly rich cloudlike formation of poems, stories, rites, performances, booklets, and videos, many of them instances of the social dance of a small group of (imaginary) characters across the surface of a world (imaginary) with very specific rules.

This continuity and interweaving occurred not only in media, genres, and idioms but across a broad range of social environments: the experimental art and music worlds, academia, nonprofit organizations, and the business world. Beginning in the late 1980s Rich's presence was increasingly in demand at research institutes and conferences concerned with everything from innovation and art, interface design, and civil engineering, to publishing, language, multicultural technology transfer, and ethics.

Using the tools of one discipline to do something other than its original intent, recalls linguistic and anthropological studies in a kind of mimicry that can be understood as "faux" science in the dadaist spirit. But the use of tools of analysis to create art and of artworks to comment upon and analyze the world is also fraught with broader significance. In "The History of Modern Golf" a tenth hole is "grown" as if it were an embryo. The conversion of one thing into another, for example a fictional or invented topography used to generate sound, the reading of one system through another—these processes, their results, and what he made out of them, Rich called *structuristems* or "Modern Sophistries."

STRUCTURALISM

Modernism had foregrounded the processes and materials of art making, calling into question the relation of form and content. Structuralism was modernism's mid-century manifestation in the world of ideas. Rooted in formalism from eastern Europe and Russia, it flared into western European thought with Levi-Strauss's uses of linguistic analytical models to study human cultural patterns. Structuralism brought into focus the strange disjuncture of Signifier from the Signified, providing ways to step back and fiddle with these relationships. Thus did the art world move toward its inevitable confrontation with the most metaphysical of relations: that between an artwork and its meaning. Did the "meaning" of an artwork pre-exist? Where was it to be found? Was it produced or constructed, and if so by whom? From Duchamp's urinal to Serrano's "Piss Christ" this urgent question has been tattooed into mass awareness.

FROM DADA TO DATABASES—PLAY AS WORK, WORK AS PLAY

Rich's roots were in artistic movements such as futurism and surrealism. But the avant garde of our 1960s youth—John Cage, David Behrman, Bob Ashley, Gordon Mumma, Alvin Lucier, Fluxus, the Art and Technology movement—were most influential. He moved through disparate even contradictory worlds of low-budget experimental art venues, arcade-game design, toy and entertainment product design, and ultimately the business and academic arenas, fully engaged with the minute tangle of present difficulties but always keeping the big picture in mind. Rich never lost the sense of *play* so crucial to the trickster and inherent in those avant-garde movements; for him, work and play were not distinct and separate spheres but part of his daily life.

Rich's first homegrown electronic entertainment, the Train Game, landed him a job at SEGA in the early 1980s, where he produced digital sounds for coin-op arcade games. He had entered the corporate umbrella world by way of games, interfaces, toys.

After Sega, Rich developed an interactive entertainment product for home computers consisting of the Little Computer Person, a "pet person" living inside a virtual dollhouse. This was years before

Tomaguchi was popular. Subsequently, at Mattel, Rich worked on projects incorporating electronics : Baby Heather, Captain Power, AI Robot, and the Power Glove. As he learned about the design, manufacture, and marketing of products, he understood that, since "play" is a highly malleable activity, the "Toy" can be viewed as an interface to the imagination. PowerPoint itself functioned for Rich as what he called a "Toy for Thought."

The Goldograph chapbooks are whimsical applications of what he called *toy structuralism*. In the early 1980s a set of elements emerged: There were ten of them, of course, so that a ten by ten grid could emerge as they "mated." The TEN OBJECTS are as silly and arbitrary as anything can be, yet, when displayed, the matrix of all 100+ images and their captions oscillates uncannily between absurdity and pathos.

After working on "ubiquitous computing" with Mark Weiser at PARC, Rich continued to focus on distributed computing as it became an everyday part of lives in the form of GPS, cell phones, and PDAs. He would use the concept of devices as *thought tokens* or EKOs, evocative knowledge objects.

His enquiry extended the observations in that 1978 *Ear* magazine article. The free-wheeling intermedia/multimedia experiments of the 1970s had taken a strange turn and everything more or less converged onto "The Computer" in its now normative familiar form.

READING AND WRITING

Another thread woven through the textile of Rich's work-life is an abiding interest in literature, literacy, and the technologies of reading and writing. "Ink must be in my blood," he used to say. His father owned a printing business that was later run by his brother; one sister is a print-maker, another had a desktop publishing business for years. He pioneered the writing of novellas in a single day in performance (sometimes directly onto walls or objects) in galleries, complete with publishing party and reading in the evening. These were not extemporaneous or "automatic" writings, but highly structured ones based on schema, algorithms, or the behavior of real-time interactive and intra-active microcomputer networks.

Some of his early works outlined the "initial conditions necessary for the rise of agriculture on a toroid world given programmatic vegetation and a single nomadic tribe," studied "infidelity as a 2-dimensional vibrating surface in a frictionless proverbial world," or used nida structures (a cross between surface and deep grammar) and its associated binary-tree diagrams to perform syntactical investigations on random sentences.

The Border was a set of performances based on a novella written using a cartoon surface structures semantics and a proscriptive syntax. His conceptual approach interleafed relationships between various types of structure. He called this "vertical" literature, where readers would descend through layers and levels, as opposed to the more horizontal traditional textual practices where the text is meant and assumed to be a transparent medium through which its signified can be absorbed.

In the emerging digital world in the early nineties, the "copier giant" Xerox repositioned itself as the Document Company. The group Rich headed at Xerox Palo Alto Research Center (PARC) eventually became known as Research in Experimental Documents. Experiments in the Future of Reading was produced by RED for the Tech Museum of Innovation in San Jose. Thirteen interactive installations explored how the technology and mechanics of reading have changed and offered a variety of approaches to ways we might read in the future.

MID-CENTURY VISIONARIES: DICK AND DEBORD

Finally, it is important to mention two mid-century visionaries who were crucial to Rich's thinking. First is the compressed precision of situationism and Guy Debord's early grasp of how mass media and technology would profoundly alter social consciousness. Debord was the initiator of this cynical and prescient Parisian form of post-modern anarchism that has successfully resisted assimilation by mass media and remained relatively obscure to this day. Rich also read and collected the books of Philip K. Dick, the California science fiction writer whose stories are still being made into chilling major movies nearly fifty years after he wrote them.

In a fury of insight during the 1950s, Dick and Debord foresaw the future and its discontents with shattering clarity. Although Rich was of a more sanguine temperament than these dystopic thinkers, he kept their concerns at hand as he wrestled with aesthetic, social, and global issues.

Within the constraints of the corporate world and high-tech design Rich continued the investigations he began in the quirky avant-garde realm, expanding their accessibility and relevance. What he came up with as an obscure and idiosyncratic artist in the seventies was now an international commodity.

He had carved out a space where he was not only able but encouraged to ask questions: The texts of two of his last published pieces, "How Smart Does Your Bed Have to Be Before You Are Afraid to

Go to Sleep in It?" and "When My Father Mows the Lawn Is He a Cyborg?" are a series of open-ended questions amounting to a metaphysics of representation.

At his untimely death in January 2003 Rich Gold left a densely packed collection of materials documenting much of his output since the early 1970s. These are now housed in the Silicon Valley Archives of Stanford University's Special Collections Library.

I. INTRODUCTION

Between 1992 and 2002 I found myself giving hundreds of talks. Most audiences think of them, I suspect, as slightly odd business presentations; but I think of them as toned-down performance art works. I have given them to sailors in the Coast Guard, to scientists at IBM, to artists at the Adelaide Arts Festival, to designers at the Aspen Design Conference, to civic leaders in Sheffield England, to the PTA in Palo Alto, to engineers at the MITRE think-tank, to academics at MIT, and to CEOs at the World Economic Forum. I've certainly racked up the frequent flyer miles. In these talks I stand before a large screen and narrate the slide that is projected onto it. A button is pressed, the image changes, and I narrate the next slide. And so on. Sometimes I spend several minutes talking about a slide; more often, I give each about thirty seconds or so. That's a lot of slides for an hour-long talk.

I call this form of literature "verbally augmented epigraphic writing." It is practiced by generals in front of colorful war maps, coaches in front of chalk-diagrammed black boards, designers with illustrated flipcharts, and art historians with 35mm slides of old masters. In the last few years, my talks have converged on a single theme: the Plenitude.

How many things are there in an average room ... say my kitchen? I can easily count a thousand, but the actual answer is fractal. Every appliance, every tool, even every food (certainly if you count pesticide residue) is compound and is composed of tens, hundreds, sometimes thousands of other things. And every day new shopping bags arrive filled with yet more things. The bags were filled at malls and supermarkets, themselves filled with millions of things. It's a lot of stuff.

Some of the stuff is called media and it's filled with transient, slightly more ethereal stuff. Some large part of the stuff in the media are words and images designed to get us to purchase the nonmediated stuff (and services to manage the stuff.) While it is true that each piece of stuff satisfies some desire, it is also true that each piece of stuff creates the need for even more stuff. Cereal demands a spoon; a TV demands a remote. The stuff coevolved and is intertwined and interdependent. Recently some of the stuff has begun to talk to other stuff directly. The kitchen utensils, like an early Disney movie, have begun to chat and dance behind our backs. And all the while, more bags of stuff keep coming in the door.

For a long time I called our culture the Junk Tribe, but that's both too pejorative and not scary enough. Instead, for reasons I will describe, I have come to call this dense, knotted ecology of humanly created stuff the Plenitude.

This is not the book of an outsider. Like most people in the Western industrialized world, I live deep inside the Plenitude. Furthermore, I have spent my (varied and satisfying) life making more stuff for the Plenitude: art stuff, game stuff, toy stuff, office equipment stuff, science stuff, and museum exhibit stuff, lots of (technically interesting, musically suspect) electronic music, and numerous novels. All this was carried out in the often oppositional contexts of art, science, design, and engineering.

If there is a thread that runs through creations of this lifetime, it is that most of the stuff I have made is an exploration of new things to do with that interesting abstraction we call *computing*. My life has been about taking these little chips of silicon and putting them in social situations where they had never been before.

While there are various theories as to how the Plenitude started, we know it mostly grows because it creates desire for more of itself. But it also grows because it is extraordinarily pleasurable to create. This book looks at both sides: creation and consumption. For me there is great pleasure and desire in both.

This book is taken from the talks I have given. I've tried to capture something of their sound, look, and feel. Clearly, something odd happens when genres jump media. In this particular case it is obvious that narrating a slide from the stage is different from captioning an image on a page. I will attempt to keep in mind that the reader is likely alone in an armchair, or in bed, or on a train, and not in an auditorium on an uncomfortable metal seat, part of a group listening to my explanations. Social groaning and laughter are different from private groaning and laughter.

The next chapter, *The Four Creative Hats*, concerns itself with art, science, design, and engineering, the four professions that collectively have created about 95 percent of the Plenitude. Following that, *Seven Patterns of Innovation* explores some common methodologies I have found useful in the creative professions. Lastly, *The Plenitude* looks at the big picture, the nature, physics, and moral stance, the future and consequences of the Plenitude. It asks, given all this, how we should act, knowing it is hard to even make a living without making more stuff.

II. THE FOUR CREATIVE HATS I'VE WORN

Creativity is highly prized in our society; it has, as the marketers say, high positive valence. If you want to compliment a mother tell her that her daughter is "very creative." If you want to praise a child, tell them that their essay was "very creative." The United States partly bases its dominance of the world on its supposed creativity (it invented rock and roll after all.) Emerging countries, even those good at manufacturing, worry that they are not creative enough. They hire American consultants to help them become so.

In this context, creativity is not just making things (factories do that), it's creating new things, things that have never existed before. From there the definition gets tricky, for certainly every kid's Crayola drawing is new and different, and every mom and dad will see it as creative (that's why we give them crayons). So we will narrow the definition a bit: Creativity is making something new that also opens up a new category, a new genre, or a new type of thing. There are other definitions of the word, there are whole academic commissions set up to find other definitions, but this is the one I am interested in: the creation of new stuff that creates new categories of new stuff. It turns out that there are different methodologies for such creation, well-worn paths that we have turned into professions, complete with unions and uniforms—let me call them *hats*.

Did I mention that, where others might observe complex shadings and infinite textures, I see the world as a cartoon? This is one of those cartoons.

artist

scientist*

*pseudo

designer

engineer

The 4 hats of creativity I have worn.

During my life I have put on and taken off four hats: artist, scientist, designer, and engineer. Sometimes I pick one up after the other like a circus clown. Occasionally I put two or more on my head simultaneously. Each one is distinct—with its own methods, world views, precedents, predecessors, dress styles, interior decor, histories, vocabularies, alliances, prejudices, tools, techniques, and demeanors. In some real way, for me, they are states of being as different as alligators and elephants. I can walk into an office and know immediately if it is a designer's office or an engineer's office. I can instantly tell an artist's loft from a scientist's lab, even if they are filled with the same digital tools. All of the hats can be creative, innovative, productive, even revolutionary in both the political and marketing sense. I also find each hat to be a hat in trouble.

My cartoon matrix of creative hats.

I usually represent the four hats of creation using this two-by-two matrix. I have been known to talk for an hour about it, digging myself deeper and deeper into the hole of unsustainable definitions.

In the end nobody likes to be put in a square in a box, particularly creative people. Particularly a cartoon box. There are many creative professions not represented directly in this two by two, for instance those of mathematician, architect, and politician. Like all personal paradigms I can fit them in, though I am equally certain that their two by twos would be different than mine—after all, mine came from a life lived and not a culture observed. The mathematician would be placed at the very top left corner of the science square. The politician would be situated between design and engineering. Architecture is a kidney-shape encompassing the central corners of art, design, and engineering. I often get people pointing to very specific points in the matrix and saying, "I'm right here." I have also heard people say, in that postmodernist tone, "But who is putting on and taking off the hats?"

SCIENCE

In the upper-right hand corner is the hat of science. This is the new-est hat of the four, but if we look at its attributes we see that it has had clear precursors in the hats of alchemy, wizardry, mathematics, and a certain kind of physical philosophy. As a precursor image you might picture a sorcerer in a robe with a pointy hat, perched on a stool in a flask-laden lab with bubbling fluids and giant books of formulae. Or Pythagoras surrounded by scrolls and compasses calculating the relations between the sides of triangles. Now picture a contemporary scientist in a white lab coat, amid a tangled lab of tubes and wires, staring at a green monitor attached to a scanning tunneling microscope, staring at, amazingly, images of atoms. The purple robe and the white lab coat are not identical, but my guess is that the medieval alchemist would easily recognize the woman in the white lab coat as one of his own.

While there is no good way to define science (or any of the hats, really) we can say, in general, that when someone wears this hat they seek to understand the basic laws of nature and to express those laws as mathematical equations. This implies many things. It implies, above all, that the wearer of the science hat believes in a nature that exists and in a nature that has laws. The difference between science and alchemy is where those laws are placed; the latter believing that they are in the supernatural, the former that they're in nature itself. But this might be no difference at all.

The idealized scientific method, which provides a generalized ethos for the hat, starts with an insight, called a theory, derived from studying previous scientific work and from personal vision. This vision part is important. Scientists become famous, much like art-ists, based on their visions. Experiments are conceived of, and then executed, that test these visions. These experiments are observed and the resulting data are reworked into a series of equations and placed within a peculiar form of literature called a paper. The paper is peer reviewed by other scientists and, one hopes, published. Once published yet other scientists may try to replicate the results. Slowly, but surely, the equations become part of the edifice of scientific belief and indistinguishable from the belief in the real world, or for that matter, truth itself. And, yes. the process seems primarily comprised of scientists. Science is not a democracy.

Given this definition scientists would seem to create only equations and papers, which would hardly qualify science as being one of the big four creative hats. And indeed for precisely this reason many scientists do not think of themselves as being in the same matrix as designers and engineers. Heaven forbid.

But the world is far messier than the ideal, for not only do we think of scientists as having created everything from rocket ships to cures for cancer, from genetically modified corn to atomic bombs, but so do the scientists themselves. In some cases these inventions were the goals of the scientific enterprise, in others they tightly intertwined with the discovery of the laws that they exemplify. While the scientist may care more deeply about the laws of nature than these physical inventions (say the cure for polio) they are both the justification for and the proof of the scientific enterprise.

People wearing the hat of science also create rooms full of very cool test equipment. This equipment is to science, I think, as the guitar is to rock and roll. Only you can't buy it at Guitar Center. There is much cleverness and invention in these beautifully engineered tools; they alone qualify the scientist for a hat in the Plenitude. It is, after all, the equations that flow from the scientist's bench that give the engineers their head-starts, their big hints, allowing them to make the goods that fill our shopping centers. Seen this way science is amazingly creative. It may be that, without science, there might be no Plenitude. But . . . this enthusiasm for new stuff threatens to overwhelm us.

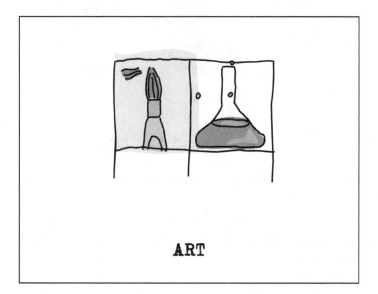

ART

I passed over the different subhats of science—for example the differences between applied and fundamental science, or the distinctions between natural and artificial science. For art, however, I am compelled to delve more deeply into subhats. That's what happens when you get a master of fine arts degree. Suddenly these sorts of things matter. I speak from the viewpoint of a recovering artist—or perhaps postartist is a better term. To my cartoon eyes, there are three kinds of art hats to be found in the Plenitude.

Fine, Pop, and Folk Art.

Fine artists, as symbolized in this slide, wear berets. These artists work from within themselves, from their visions. They try to express themselves and their ideas, and the resulting art is a representation of those ideas. Beret wearers seek a kind of truth (they often use the word integrity) where the art they produce is equal to the visions. Compromising the vision is seen as a form of evil. They study the world and art history and immerse themselves in an ongoing, esoteric dialogue that chatters away unbeknownst to most people. This process generates the art you see in museums, galleries, books, magazines, and mansions.

When this method works, when the beret is really cooking, the work has a semimagical, transcendent quality about it. Just as alchemists would recognize scientists, shamans would recognize the fetishes of contemporary art (and vice versa). Art often exists under separate laws from the other stuff of the Plenitude. For instance, in many states it is illegal to destroy a work of fine art, one assumes because, like scientific equations, they are intended to have a validity beyond the moment. There have even been serious moral debates about whether soldiers should risk death trying to save works of fine art. Or whether a public sculpture that inconveniences or offends some people can be removed. Or whether it is right for one country to own the art works of another. Not the sort of bottom-line, branding questions that arise when talking about, say, Swatch watches.

The basic economics of the fine art beret goes like this: Those who wear it make their living (if they make a living) by producing a small number of objects that they sell for large amounts of money, usually to the corporations, governments or the wealthy. In the not-so-distant past they were also funded by the Church. That is, fine artists are supported by the most powerful, elitist, influential forces in our culture. Oddly, of course, the beret wearers often see themselves as outsiders, or even antithetical to the power elite. It is traditional for artists to see themselves as agents of change attempting to bring down, or at least *alter,* or at least *awaken,* or at least *offend,* the ruling structures. But while articles about artists in the front of art magazines present a new world, the gallery ads in the back appeal to the old, established world. Precisely because art transcends, it becomes valuable as an economic hedge against all forms of calamity.

To accomplish this remarkable sales feat—selling paint on cloth, interactive gizmos, or whatever for large sums of money—artists must convince patrons that they, at least when wearing the beret, are of a special nature and embody a unique process. They are not like accountants who could have chosen to be dentists instead. It's not uncommon for artists to claim that they were *born* artists, that they have always been artists, that they can't help but be an artist. From inside this special self emerges an idea or vision that when realized, well or poorly, is the work. These works are evaluated by the artist's peers (and near-peers including curators and critics) and are judged good, or bad, or coyly interesting. It is not democratic. While adventurous patrons may buy an individual work or two, it quickly becomes the artist and his or her oeuvre as a whole that becomes valuable, the works becoming only exemplars of the process and the vision. And it is only when patrons begin buying the artist, however long that might take, that the artist and the data points of their work becomes part of the Plenitude's art edifice. Sometimes the trust fund runs out first.

With the beret of fine art on, a creator looks within for inspiration. The art flows from personal vision and from a unique sense of self. To many artists, art is more a calling than a profession, though one still needs to be trained in it, and there is certainly a business side. While the art vision flows from within the artist, what the artist is representing are certain deep aspects of the world. These can be from nature or human nature, or they can be cultural, autobiographical, linguistic, historical, political, or even religious. And these artistic insights are intended to be deep, for art produced in this fashion is intended to last millennia. One of the prime values to the owner, the patron who buys it, is the monetary benefit that accrues to unique objects of long-lasting value. Kind of like manmade diamonds. In one sense, then, artists are like the scientists, looking for, dare I say it, Truths, even if only personal ones. As in science, the work exists within a complex dialogue with peers, a hermetic language pretty much impenetrable by the average citizen. One difference from the science world, however, is that works need to be unique from artist to artist. The value is much about the uniqueness of the individual. Replicating art does not "prove" it.

Stuff made with the art hat on can be enjoyable, insightful, funny, ironic, beautiful, entertaining, enlightening, inspirational, soul wrenching. I have been almost brought to my knees by a few pieces. There is, of course, way too much of it. One might ask what art colleges think they are doing graduating that many art students. Nonetheless, it is an extraordinarily useful hat to put on from time to time, even for the nonartist. The hats of design and engineering need the deep infusions of vision that the beret can provide. Without artistic vision stuff tends to asymptote to commodity. Lesson for corporations: If there isn't a little art in what you do, the kids will wander off to buy somebody else's sneakers.

Many artists believe that wearing the art beret is antithetical to the Plenitude, maybe even its antidote. Or at least a refuge from it. They believe that art's close ties to the ruling classes are just a ruse, a trick pulled on the patron to get money. This is not my belief. I believe that Western fine art is almost a perfect reflection of the society that produces it. From the love of the new, to the cult of the individual, from the commodification of the aesthetic surface, to the elaborate laws of intellectual ownership, from the concepts of continual revolution and change, to the belief in modernity and postmodernity—art and society are strange and perfect twins.

There is even an argument that our civilization's base capitalistic concepts first arose in the art world and were appropriated by the rising bourgeoisie; that the Plenitude is what you get when you cross art with the corporation; that if we're heading toward global destruction, the principles of Western art are not the life preservers but the ship that's taking us down! *Note to self: This is a flimsy argument.*

Second, there is the popular art hat, represented by the baseball cap. The cap-wearing popular artists focus less on their inner vision than on the emotions of their audiences. They, or their producers, managers, and agents, follow the top 100 charts with a magnifying lens; they study the box office numbers like accountants. They will do, and this is a major distinction from beret wearers, user-testing! To the beret wearer this would be an outrage, for it would be the loss of integrity, not to mention vision. But not to the true baseball cap: The deep feedback loops between themselves and the audience are as important as the work itself. In this sense, the baseball cap is democratic, antielitist, and human. The idea of producing art that confuses or repels is deeply offensive to the baseball-cap artist.

The economics of baseball-cap art is to make highly replicable works—movies, pop songs, television, video games, clothing, and most books. The works themselves are often very expensive to produce (a movie, for instance, might run to hundreds of millions of dollars) but the goal is to sell an enormous number of them for a relatively small amount of money. The artists are not supported by corporations so much as they are part of the corporations, and they rely on the corporation for manufacturing, advertising, and distribution.

The baseball-cap artistic hat probably could not have existed much before the mid-nineteenth century. Today popular art constitutes a large percentage of the Plenitude both in the number of individual works and the work's multiples, which run into the billions. At its best, it is world encompassing and far more inclusive than most other art. It is the art type most indicative of our time. In a thousand years it's unclear whether the Beatles or Steve Reich will be remembered more clearly, but my guess is that it will be the Beatles. Perhaps to future listeners they will be indistinguishable.

At its worst, the audience-artist feedback loop that sits at the heart of popular art can produce a kind of mind-numbing, full surround cacophony that constitutes a form of pollution. When my fleeting, trivial desires are amplified by focus groups and blared back out through multiple channels, even the car radio can become an unbearable torture.

Lastly, are those artists we might call folk artists, symbolized by the straw hat of the banjo strummer. This is a particularly unfair image, I must admit, for folk art ranges from people who make furniture in their basements to teenagers making rock and roll in their garages. Here's my definition: Straw-hat artists make art for themselves and for their friends; they engage in art making not because it will last forever, or because it will please a million people, but because it is fun, enjoyable, and satisfying; because it is a way of interacting with and strengthening the bonds between friends and family.

Until the first great washes of the Plenitude crashed over the populace, what we now call folk art was simply called music, or dance, or drawing and was the vast majority of art produced. It is still produced in some quantity, though it gets drawn like a magnet either to popular art (the rock band in the garage almost immediately thinks about selling CDs) or to fine art (the weekend painter almost as a reflex imagines his or her work in a gallery.) Or a Martha Stewart figure turns it into a kind of user-configurable corporate product. The inability to make art simply for its own enjoyment is one of the great losses, let me go further, is one of the great tragedies, of the Plenitude.

The economics of the straw hat of art are this: While it costs something to produce it, there is rarely money exchanged, or at most, only small sums between the maker and the consumer. It is, as the anthropologists say, a gift culture and it is to a large extent gone. Hakim Bey calls it *immediatism*, something that is done for immediate pleasure and for the immediate company. In our society fantasies of making money quickly overpower such activities. Another reason is time. There is never enough time, and in a crunch out goes the folk art. Q: What takes up the time? A: The rest of the Plenitude.

For instance, it is fairly rare these days to go to somebody's house and have them play music for you. Yet only a hundred years ago this was the most common form of music. We further denigrate such activity by using such phrases as weekend painter or vanity-press author. Straw-hat street forms, such as rap and break dancing, get so rapidly pulled into baseball cap art that by the time it reaches the Minnesota suburbs six months later it is already a popular art and not a folk-art form. Even home cooking, a true folk-art form, has been greatly replaced by home-*style* meals you can buy in a supermarket.

In densely worded papers, contemporary cultural theorists like to claim that folk art still exists at the edges of popular art—from making new houses for The Sims, to writing Kirk/Spock homoerotic stories (and posting them on the web), to remixing pop songs, to recutting Star Wars movies, to inventing strange rituals in Ever-Quest. It hard to argue with this, but it is also hard not to be a little cynical, particularly when the companies producing these pop art works are banking on such folk interests. I do believe that the truly beautiful explosion of graphic arts in the mid-1990s that occurred on the web was mostly folk art. It certainly didn't make money. It was just immediate fun to do for you and your unmet friends.

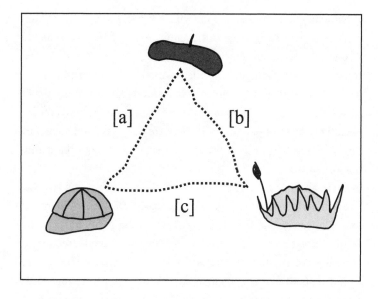

A more legitimate way of presenting the three hats of art is to say that they form a triangle and that any given artist (or work) can be placed somewhere in that triangle. Some artists are very close to one vertex or the other but most are somewhere in the middle. It might be useful to look at some of the artistic types that exist on the edges. Here some data points:

Rock and roll is generally thought to be a popular art form with its genesis in the folk art forms of blues and country; Dick Dale says that his surf music was the combination of rock and Lebanese folk music. There are fine artists who experiment with rock and roll (e.g., Laurie Anderson) and pop stars who toy with fine art (The Beatles' "Revolution #9"). Fine art composers have often drawn on folk art (Stravinsky and Bartok come to mind); much piano playing at home is actually of Bach; Bob Dylan rewrote folk music and then his songs were sung around the camp fire as if they were as old as the hills. There are more unusual cases, for instance, there is a sense in which young cutting-edge painters are folk artists. Some large percentage of them the are sons and daughters of the upper classes of our society. If their works are purchased they are bought, essentially, by their parents and if they are not, they were done only for the enjoyment of their friends. Together, just about the definition of folk art.

There are many similarities between the hats of art and science. They share a common progenitor: Leonardo Da Vinci. When one listens to artists and scientists speak one hears many of the same words, though often with different shadings or even reversed meanings. But the territory, the frame, is the same. They both speak of the future (and claim to precurse it); they both speak of truth (sometimes about finding it, sometimes about denying it); and they both talk about saving the world, sometimes from each other.

Both artists and scientists talk about nature. For the scientist it is a central object of study. Artists ping all around the concept—from postmodern denial of its existence, to making it the central theme of their work. It is, as they say in the lofts, a site of contention. Both artists and scientists speak of personal vision, of having and working from visions, which are different than mere ideas. They think of themselves as unique (for the artist this is the bankable commodity.) They both speak, in other words, in transcendent signifiers, giant words that by their nature are outside the world. Both hats contain a concept of genius, what the Nobel prize is all about, and how both Picasso and Hawking are referred to. The art community might make noises about there being "an artist in all of us," but try telling that to curators, dealers, and acquisitions committees of museums—all artistic homunculi are not equal. There may be talk of the "little scientist" in each child, but the adult scientific community (not without cause) has rigorous entrance standards.

DESIGN

In the lower-left hand corner is the hat of design. As I travel around the country presenting this little 2-by-2 matrix I have found that that many corporate executives don't distinguish between those who wear the hat of art and those who wear the hat of design. They certainly don't know that when the hats are tightly worn, the wearers hate each other. Maybe hate is too strong a word; we might say that the methods represented by the two hats are more or less diametrically opposed to each other and cause procedural friction. Woe to those who mistake an artist for a designer and ask for a little more green in the painting to match, say, the couch. And woe to the designer who forgoes the opinions, desires, and needs of his users and clients to rely only on his own visions and dreams.

For an artist user-testing is a joke. For a designer it is fundamental. If an artist looks inward as a way of seeing the world, the designer looks outward toward others. An artist paints a painting, stares at it, and says, "isn't it beautiful, it expresses my inner vision perfectly." The designer paints a painting, stares at, then turns it around to the audience and asks "Do you like it? No? Then I'll change it." When it works, when the designer can hone in on the audience's wave length, it is an amazing and beautiful trick. It has created most of the bounty around us.

All cultures are designed, of course. It is through design that we can distinguish the arrowheads of this tribe from that; the houses of this civilization from that one; the utensils of one ancient nation from the next. But the hat of design, as we think of it today, separated out as it is from other employment, is new. You can actually get a degree in design and figure out what you will design later on. Should I design dolls or websites? Cars or advertisements? It is the story of my life.

The designer speaks a language quite different from that of an artist. *Precision* is a word designers often use, as is *brand* and the ubiquitous, ambiguous *user*. Design that doesn't communicate with the user, or satisfy the user's needs, is considered poor design. Design that is messy at the edges, or that gets in the way, or that needs to be fought with (as art often requires) is considered poor design. Design is functional and usually serves a purpose, even if it is simply to entertain, beautify, or sell something. It can be used to communicate, make something easier or more comfortable. It is often used simply to say, "I care about you."

The wearer of the design hat often thinks of the wearer of the artist hat as a navel gazer since the viewer is ignored, or even ridiculed; while the artist thinks of the designer as somebody who sold out; who couldn't make it in the semimystical realm of art. There is much less transcendent signification in design than in art. It is not above the world, it is of it.

In our culture the artist is more revered than the designer (they teach art, and not design, in grade school.) But it is the designer who has had the much greater influence within the Plenitude. When design is revered it is usually the design and not the designer that is honored, for unlike the artist, the designer is often anonymous and can be switched out at the whim of his employer. Oh, did I mention that designers have employers?

This modern hat of design flows, in great measure, from a peculiar aspect of the Plenitude that is so important and vital that it is actually embedded in law. Every new thing, every new hunk of stuff, must be different from every other thing. In many cultures, when you tell a story, that story is the same as the story your parents told you. And that their parents told them. Every knife you fashion is just like the knife the your elders fashioned, and their elders. But in the Plenitude, this is illegal! It is called copyright infringement! In the Plenitude every story, every novel, every movie, every knife, every car, every cereal, every toy, every napkin, every napkin holder must be different than every other. And it must be different this year from last year. It must be new and improved continuously, and if the oatmeal itself can't be improved, then the packaging must be.

If art looks at the deep fundamentals that should last five hundred years then design looks at the trends that must be novel this month, and need only last this season. Design looks at fashion. For the designer, in a real sense, when the car is sold, their job is done. The cumulative effect is to produce the glory of the Plenitude, which is not so transitory. The designerly Plenitude, as a whole, is our culture's greatest art.

ENGINEERING

In the lower right corner is the last of the four hats of creation. Engineering is the hat of problem solving, of rules of thumb, of simple machines (pulleys and gears) and complex ones (flying buttresses and rotary engines), of numerical tables, of equations, and of books of regulations. It is "necessity is the mother of invention" and "do no harm" (first of all, the bridge should not collapse).

The things that engineers build are bounded by constraints, from the laws of nature to the laws congress imposes on fuel mileage. The job of the engineer is to get the world of molecules to act in ways that will solve engineering problems. People need light at night: Now there's a clear problem. That's an understandable desire. Engineers want their problems in the form of a clear spec, so they know when they have succeeded. Engineers believe that within the fixed bounds of the laws of nature, there is the solution to almost every problem. Finding it is the job. Once we wanted to fly, we flew. Once we wanted to get to the moon, we did. And now that we want warp drives, we will engineer it. To the engineer it is not whether it is solvable or not, it is whether it is a hard or an easy problem. They differ by how long it will take to find the answer (and how much money it will require.)

The hat of engineering is closely related to the hat of design. Both work from need and desire. Both are concerned primarily with the user and the world—the "real world," as they like to say. Unfortunately, in most companies design is pitted against engineering, a battle that tends to reduce the effectiveness of both. I think this is caused by a misunderstanding by both engineers and management, who see the hat of design as the hat of art. They think that designers work from inner vision and not problem solving.

Engineering's relationship to science is more complex. Engineering is the oldest of the four creative hats. Egyptian or Roman engineers (pyramids and aqueducts) are almost indistinguishable from contemporary ones in their methodologies and basic world view. Of course their tools have changed, as have their books of formulas, rules of thumb, exemplars, and equations. But these are always on the march. Engineering, without science, might well have created the entire Plenitude, though a somewhat different one and it would have taken longer. From the cathedrals of Europe to the airplane and the car, from Chinese porcelain to the sailing ships of Europe, from the ziggurats of Persia to the cultivation of grain, all were created by people wearing the hat of engineering, not the hat of science.

Science, to the engineer, does two things: First, it presents new equations about how the world works that enable the finding of engineering solutions. Second, and more interestingly, science creates new desires and needs that engineers must then solve. Science fiction rests on new science, but the worlds presented in science fiction are pre-artifacts that the engineers will create.

While science is a hat of laws, engineering is a hat of violations. Engineers spend a lot of their time engineering out the exceptions. Most of the stuff in a high-speed printer is there to take care of the jam that occurs only once in every thousand pieces of paper. Much of why a car is built the way it is is to prevent or minimize the accident that will occur only once or twice (or never) in its lifetime. There are those who believe that all of science will eventually be a single, beautiful equation. Not so for engineering. To the engineer all is local, unique, different, problematic, ready to violate every known law. The engineering library expands constantly. Engineering forms the backbone of invention of the Plenitude. If you want to make sure your child has a job in the future, have him or her become an engineer. If you go back into the past using a time machine and you want to insure your survival, bring along an engineer, not a scientist.

Design and engineering are much more related to each other than art is to science. It's not good design or engineering unless other humans like it, buy it, use it. It is irrelevant whether it satisfies its creator's vision. Design and engineering create the physical artifacts by which we interact and communicate with one another. Your clothing is a visual language that is perpetually speaking; your car is a socially negotiated piece of metal to get you from one human to another. We might treasure a design/engineering effort from two hundred years ago, but it was created for a moment. What is good design/engineering today could be bad tomorrow. Good design in one part of town could be bad in another. Engineered crops in this country could be (and are) illegal in another. Design and engineering are rhetorical devices in the sphere of human exchange. They form the physical language a culture speaks in a dialogue about everything from how we will house the elderly to how we shake the salt.

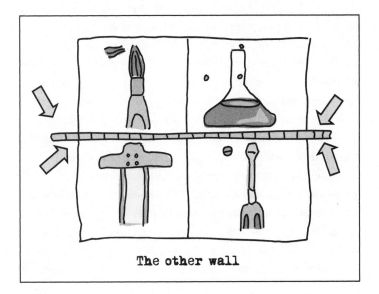

The other wall

Most people think that the largest schism in our culture is between the arts and the sciences, as expressed by C. P. Snow's *Two Cultures*. But in my cartoon model, the biggest schism is between art and science at the top, and design and engineering at the bottom. The horizontal line is called the Wall and it is notoriously difficult to get something over it. It is difficult for the scientist to hand off his or her work to the engineer. There is a similar wall in the design world. For instance, a truly great toy inventor is an artist: the toy comes in a vision and is pursued in that way visions are pursued. But when a company takes it on as a toy to manufacture it must go over the wall where, to the horror of the artist-inventor, it gets mangled by a committee of in-house designers. It's a war. Our culture is divided in two, but not as Snow thought. It is divided horizontally by the Wall.

All professions have "others"—other professions or groups to which they are deeply related. One of the big differences between art/science and design/engineering are their *others*.

In art and science the others are the Patron and the Peer. In the science world, the peer is a built-in and overtly expressed part of the process called *peer review*. A scientific truth only becomes true when one's peers (not one's neighbors or the congress or a company) agree that it is true. For the artist the peer group is other artists and the related profession of the critic. The patron is the means by which artists and scientists get funding. In today's world such funding usually flows from governments, businesses, or the rich. Universities also figure in here as the Church used to. Patrons usually care more about the artist or scientist than the specific art or science they are doing. That part is up to the vision of the artist and scientist. What's particularly odd is that while the patrons are the most powerful institutions on the planet, artists and scientists often see themselves as outsiders. There is something almost pathological about antiestablishment artists trying to have theirwork presented in patron-sponsored museums.

For design and engineering the others are the User and the Client. The client presents the designer or engineer with the problem and pays them a fee to solve it; the client gives the little red checkmarks saying the job has been done, the spec has been satisfied. The user is the person who actually engages with the artifact that has been designed or engineered. For instance, in the graphic arts, the client is the person who orders a poster to be designed, who says its OK or that the letters should be bigger, or whatever, and who pays the money when it's done. The user is the person who sees the poster on the wall and, one hopes, acts on it by going to the concert. There is much debate in these communities between client-centric vs. user-centric design/engineering, but in either case, the interactions between the user and client are important and can produce wondrously delicious ideas.

Design without art, or engineering without science, both quickly asymptote to commodity, and in the globalized world, if you are merely producing commodity, you're dead. Learning to cross the wall and making peace between the peers, patrons, users, and clients is the most important lesson I know of for a corporation.

But, please remember, these are just cartoons. It is how I see the world, not how it is.

For many years I ran a program at Xerox PARC, a scientific research center, that brought in artists to work with scientists. The program was called PAIR (PARC Artist in Residence) and it was quite successful, perhaps because artists and scientists are very similar. I then brought in designers to work with the scientists. This was much less successful. I believe that it was because scientists thought of designers as very close to marketing and that marketing is about lying, while science is about truth. There could be no larger difference. And so, while yes, each of us spans all four cells of the matrix, they are not the same. You will seldom mistake an art conference for an engineering one. The designer's studio is rarely confused with a scientist's lab.

Though having said this, it is just these confusions I find most interesting.

And it's been the story of my life.

III. SEVEN PATTERNS OF INNOVATION

As I bounced from quadrant to quadrant among the four professional arenas of art, science, design, and engineering, I found there was actually a shared and limited set of methodologies for the creation of new stuff. These seven methods or patterns are used in all four disciplines, but are differently weighted in each area. These weights, at least in part, determine the quadrant's soul.

Each pattern also comes with at least one danger; a kind of counterforce that renders it difficult to harness, or dangerous to use. The Plenitude arises from the interaction of these patterns and forces. As with DNA, there is enough complexity so that the products are unpredictable and enough simplicity that they are understandable (at least in hindsight).

Together these seven patterns and their counterforces create a kind of field-effect. Most designers, artists, engineers, and scientists don't think of them as separate little magnets pulling and pushing this way and that but as an encompassing creative field. You just wake up in the morning and start working.

So why specify them? Here's one use: Every so often you come to your desk, your workbench, your design table, and you are stuck. No ideas flow. Everything seems already invented that needed to be invented. It is at that moment quite helpful to pull out this little list. I have also found it enlightening in the other direction. I see a new product that I want to comprehend. How did it come to be part of the Plenitude? I can hold the product up against the grid of these patterns and understand, if not how it was born, at least how it was patted into shape by these forces of creation.

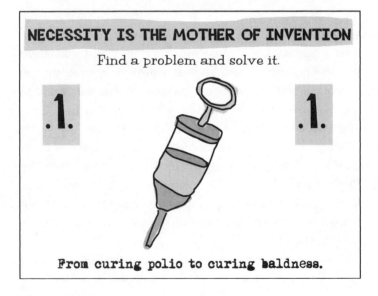

NECESSITY IS THE MOTHER OF INVENTION

Find a problem and solve it.

.1.   .1.

From curing polio to curing baldness.

The first pattern is one that most people will think of off the top of their heads. It goes like this: There are a whole bunch of problems out there and our job is to find these problems and solve them one after another. When you're in college training to be an innovator these problems are big: cure cancer, solve world hunger, create peace, change history with a novel. Of course, when you get your first job it turns out you're going to help people read their e-mail better, assist them with their hair loss problem, or write programs to help them remember birthdays. Yet, small as they may be, they are still real problems and people will pay you to have them removed. It is good honest work. If you can't think of anything else to do, it is usually worthwhile to find a problem and invent a way to eliminate it.

This is a well-known pattern and there are many good books written about it. It has one primary defect, however, that can be simply stated. Every solution to every problem creates ten more problems.

In 1900, the cities of the United States had a huge problem. It was called horse manure. It wasn't the speed or the cost of horses that was the problem. Horses were actually a pretty nice way to move around a city. It was the manure that was the problem. Horseless carriages, what we now call cars, were invented to solve the manure problem. And the solution worked. But, in the process, they created air pollution, gridlock, the destruction of the center of our cities, parking lots, reliance on foreign oil, global warming, war, and, possibly, the end of the world. The problem was narrow, but the consequences of the solution were wide.

There is another kind of trouble that can arise from the simple use of this pattern. It happens when you attempt to solve each individual problem separately. For instance, a radio has all these little problems: You can't remember your favorite station, you need a different treble and bass setting for different songs, you need a remote control, and so on. Hundreds of little problems, each of which can be solved simply with a button. Each solution works fine. But pretty soon you have this radio with fifty buttons on it and it has become so complex you can't figure out how to turn the thing on. It can be a dangerous pattern to use. But as you can tell from the Plenitude we live in, it' s the easiest one to get paid for.

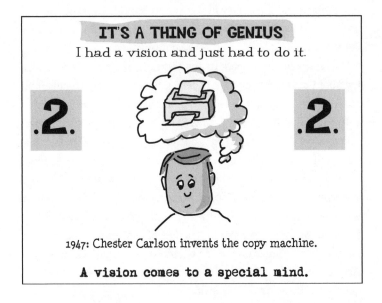

IT'S A THING OF GENIUS
I had a vision and just had to do it.

1947: Chester Carlson invents the copy machine.

A vision comes to a special mind.

While making coffee or taking a shower, while staring at a blank canvas or pounding at a keyboard, visions simply appear in the mind of the innovator. Wonderful visions, so powerful that the innovator has to do them, is compelled to realize them. They are special visions and in many ways above critique or even reason. It is considered a high moral act to follow these visions; it is considered immoral, illegal, even crazy to destroy something that came from such visions (think of burning a Picasso). This pattern relies on two preconditions: First is the belief in a special mind, the mind of genius. This genius can be inherent, or it can be the result of practice and training, usually both. Second, it relies on a deep personal belief in one's own visions that compels the innovator to make them real.

In 1938, Chester Carlson had a vision and it would take 20 years of dogged pursuit before somebody would buy it. In fact, for most of those twenty years, when he was alone with his vision, people said it was a bad idea. "Who needs such an expensive machine? We already have carbon paper." But Carlson did not give up. He believed in his vision. It came from a mental process that he trusted. It did not come from the world, except to the degree that the world was processed through his gray matter. The copy machine seemed to solve no known problem. There was no necessity that forced the invention. It was, quite simply, a vision. The *Mona Lisa* also was not the answer to a problem, either, and it too might have failed user testing. Visions are *not* about solving problems.

But here is the weird thing, the unexpected thing. People like products that come from visions. In many cases truly new things only come from visions, they don't appear step-wise by solving one problem after the other. And we have a special feeling toward products that come from visions. We put them in museums, for instance. Collectors give millions of dollars to possess them. Entire industries are founded on them. The Barbie doll was another product that was the result of a vision. There was no problem that Barbie solved. There was no clamor for a busty doll. But when Ruth Handler saw the European prototype, she *knew*, she had a vision.

If you're a corporation it seems to make sense to devote some of your energy toward creating things of vision. Xerox PARC was set up around this model. Collect a bunch of geniuses, put them up on a hill in a white building, separate them from all the problems that might be out there, and see what visions they will have. But, there are some well-known problems with geniuses.

Problem One: It's hard to know what to do with a vision. Problem Two: They can get genius block and then you're stuck with somebody with no new ideas and a high salary. Problem Three: It turns out that, if you collect a whole bunch of geniuses together in one place, the visions don't multiply, very often they subtract. Problem Four: A company can build too much of an ivory tower. While visions don't flow directly from problems, they often grow, like a pearl, from an irritating grain of sand. Lastly: It's impossible to tell the difference between a crackpot and a genius until visions succeeds. Until 1959, nobody knew if Chester Carlson was a crackpot or a genius. Success is only relative to the world. Released too early it can fail; released too late—well, then it is too late.

These are the dilemmas of corporate research centers to which genius has been relegated, ghettoes far from where the work, as the rest of the company sees it, really gets done.

THE BIG KAHUNA

Scientific deduction of stuff from 1st principles.

.3.

.3.

Ubi-Comp and the Ubi-Man.

The third pattern is something I continue to refer to as the Big Kahuna despite its unfortunate association with sixties surf movies. The Big Kahuna is the derivation of new stuff from first principles. From BIG giant ideas. From laws of the universe (or at least laws of human nature.) From inclusive theories of everything. From induction. As if induction were real.

Ubiquitous Computing, the research program I worked on at Xerox PARC, was based on a Big Kahuna. Dr. Mark Weiser first formulated it. His Big Kahuna went something like this: In the future, networked computation will exist and it will be tacitly, universally, and invisibly embedded in the everyday objects of the world. It is a wonderful vision, and I believe it will eventually will come to pass. But what this Big Kahuna did not come with, what no Big Kahuna comes with, are the details. How does this idea, this force, this direction, get realized? The Ubi-desk. The Ubi-chair. The Ubi-paper. The Ubi-shirt. The Ubi-pen. The Ubi-man? In a way, the Big Kahuna sets up an alternative set of universal laws from which one creates and invents. It is an invention that presupposes more invention.

While the Big Kahuna is a particularly strong and compelling way to create new products for the Plenitude, it has an associated series of problems. Our culture, perhaps any culture, is defined by a *collection of interlocking and often contradictory* Big Kahunas. Any attempt to create product based on just one sends you into immediate collision with the others.

When the ubiquitous computing team created The PARC Tab, the design was based on the principles laid out in Weiser's Big Kahuna. The Tab, the proto-Palm Pilot, was like a pad of paper intended to tacitly disappear into the daily work environment. But there was another Big Kahuna at play: Humans will communicate with other humans if possible. What the Tab really did was allow e-mail to be read at meetings, and so, instead of invisibly helping meetings it totally disrupted them. Meetings collapsed in a frenzied clicking of message reading.

The PARC Badge was another ubi-invention. We thought of it as simply allowing ubi-objects to know where people were. This was not what the many journalists who come to look at it saw. They saw only the literary Kahuna usually referred to as Big Brother, and that is what their headlines all blared.

People are not like molecules, and so Big Kahunas, which seem so much like laws of nature, don't work as well as the laws of physics. It's not that they don't work at all, it's just that they are subject to the same winds of change as the rest of the culture. Of the Plenitude.

Here's an example from the toy world. When I first entered that strange world in the 1980s it was well known, it was a given, it was a Big Kahuna, that soft baby dolls don't sell to young girls. At that time all baby dolls then were made of hard plastic. Soft, cloth dolls, seemed old fashioned, from another era. No self-respecting mom would give her modern daughter a stuffed doll. There was something perhaps unsanitary about it. Partly because of this law Mattel turned down the cute and very soft Cabbage Patch doll when it was offered to them. They said, "No, no, it's a soft doll. They won't sell. It's well-known they won't sell." They even tested it and sure enough little girls wanted hard dolls. Another company bought Cabbage Patch and . . . well, it sold through the roof. Two years later, if you did market testing, only soft dolls would sell big.

And before Barney it was a well-known Kahuna that only boys liked dinosaurs.

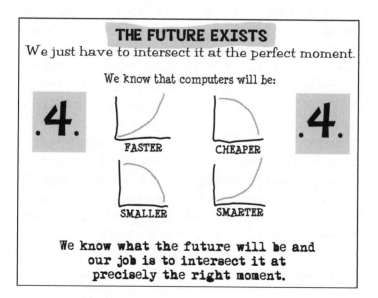

THE FUTURE EXISTS

We just have to intersect it at the perfect moment.

We know that computers will be:

.4.

FASTER

CHEAPER

SMALLER

SMARTER

.4.

We know what the future will be and our job is to intersect it at precisely the right moment.

The Future Exists. This is a pattern that you will find manifested and spoken about in engineering and research departments all over the world. Sometimes it is so strongly put that it almost has an otherworldliness about it—that the future actually already exists. That the job of the designer or engineer is to simply intersect that future at exactly the right moment. Neither too early or too late. That the future itself is a given, it is simply not yet realized.

For instance engineers know that in the future computers will be faster, smaller, and smarter. It is known that there will be more transistors per millimeter of chip, which is sometimes known as Moore's Law. There is little debate on these things; the only debate is precisely when they will occur and what technology they will use. And who will make the big bucks.

This is actually a very valuable pattern. It works so much of the time. I don't want to minimize it. It allows certain interesting methodologies. One is called buying into the future. It turns out that technologies of the future usually exist today except they are extremely expensive. If you want to design desktop software for the future you can buy a supercomputer today to do your invention work on. Five years from now, the power of the supercomputer will be on everybody's desk and no bigger than a pencil sharpener.

The problem with this pattern is simple. The future doesn't actually yet exist. It's a story that we tell ourselves to help us get up in the morning and moving in the right direction. Events can rapidly alter that future, particularly ones that flow from the Thing of Genius pattern. When we designed the first ubiquitous computing artifacts, for instance, there was no World Wide Web. Its sudden appearance made our work look as if it was heading in the wrong direction, as if it was following the wrong story. And so we created a new story about the past and now it fits in perfectly. It is not just the future that doesn't exist; neither does the past.

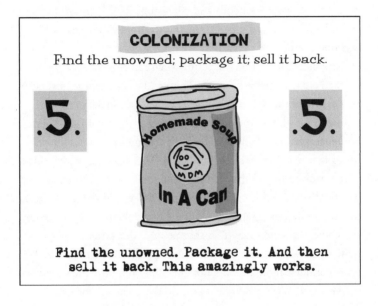

COLONIZATION

Find the unowned; package it; sell it back.

.5.   .5.

Homemade Soup
In A Can

Find the unowned. Package it. And then sell it back. This amazingly works.

The fifth pattern of innovation is the perhaps the hardest one to come to terms with. When I give talks at colleges on this topic there are often audible gasps from the young innovators in training. I call the pattern *colonization* after the Situationists International's use of the word. The Situationists were a small, radical group of French philosophers who saw the Plentitude as a massive civilization-wide spectacle. Here's how you innovate using colonization: First, you look around the world and you find things that are unowned, or barely owned, or lightly owned. Surprisingly, even in the Plentitude, in the Spectacle, this is possible. Second, you bring these unowned things into the corporation and into the maws of design and engineering groups. There you improve it, rationalize it, package it, gussy it up, make it smooth and consistent. Third, amazingly, you sell it back to the same people who once used it, for free, while it was unowned.

Campbell's Home Cookin' Soup, which can be found at any supermarket, is a reasonable example. This soup is not made at anybody's home, believe me. It is made in what we would consider to be almost the opposite of a home. It is made in a giant automated factory. Somebody at Campbell's was looking around for new product and noted that people make soup at home. They saw that people liked homemade soup and they felt it was a good thing. "Hmm, what if we could sell them homemade soup? They wouldn't have to go to the trouble of making it, and of course we would make some money. We could mass produce it so that people who never would make homemade soup can have some. We will make the Plenitude, richer, more complex, more human, homier." I'm quite sure the soup does sell very well or it would not appear on those shelves for long.

Of course, Campbell's Home Cookin' Soup is not homemade soup. That does not imply that it is bad soup (I eat it, actually). It does save time and it is healthier than eating a bag of potato chips. But you can actually still make homemade soup for free. It fills the kitchen with wonderful smells. Every bowl is different, adding variety to life. The vegetables are fresher and hence healthier. There are fewer added ingredients of the chemical kind. You can make it with a younger member of your family creating family bonds. So why care that it now exists in two forms: the free form and in its Plenitudinous form? Something of essence is lost with colonization. What makes real homemade soup wonderful is the variation and the work. For those who are still cringing, I will say that we must ask how can we make sure that as much is added as is taken away in this process.

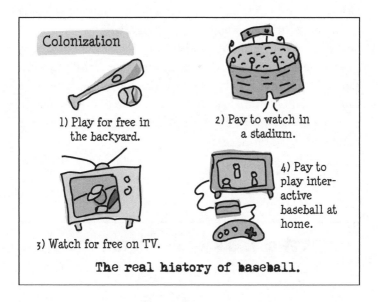

It was a magazine article that alerted me to the colonization of soup. It was my wife who pointed out the strange odyssey of the colonization of baseball. Baseball, like most sports, was created, and thrived, because it was fun to play. It is free to play once you have a bat, a ball, and gloves. You can just go out into the backyard and choose up teams. Its enjoyable and healthy. It also turns out that it was fun to watch other people play. At some point companies started to charge for watching games and "I'm going to a ball game" came to mean

I am going to go and watch a ball game being played. Colonization had begun.

Then oddly, because of advertising, baseball began being broadcast on television, where you could watch it for free without even having to leave your living room. Not only were you not swinging the bat yourself, or throwing the ball yourself, but you weren't even near other people in this otherwise highly social game. The only cost, as the web folks now say, was allowing your eyeballs to watch some ads. At his point it became clear that something pretty large had been left out of the pot. So the video game industry stepped in to colonize colonized TV baseball by making it interactive and charging $40 per cartridge.

But, you can still go in to the backyard and play real interactive baseball for free!

The sixth pattern is important, though not easy to believe, particularly for the noninnovator. For those who have never sat at an inventor's bench it can seem almost mystical. The pattern is this: Stuff Desires to Be Better Stuff.

As a designer or an engineer, you're often presented with a product that already exists (in the Plentitude) and your job is to make next year's version. Or sometimes it's not even part of your job. You're just looking around and you see some stuff. But what you really see is what that stuff should be. Or perhaps you see what the stuff itself wants to be. You will hear engineers say something like: "I know what this needs." By realizing the stuff's own needs, the innovator helps to make that stuff better. This process makes the history of any product look a lot like biological evolution—as if each year's model of car gave birth to next year's slightly evolved model.

Good designers and engineers open themselves up and let the stuff around them rant on about what it wants to be. What is really strange is how consistent stuff's desires are. For instance, all dolls want to be talking dolls. If a new doll is introduced and is successful, then three years later there will be a talking version. It was the doll who told the designer that this is what it should be. The engineers and designers working on them only make it happen.

But here's a corollary that is highly relevant if you work in a corporation: Technology Desires to Be Product. As an engineer in a corporation the problem often presented to you is not that there is such and such problem in the world; rather it is we have such and such technology and it desires to be a product. Artists also sometimes feel this desirous pull. They will see a new material, or an image, or a device, or a piece of driftwood and they can hear it calling out to be made into art. It is a wonderful feeling.

The problem with this pattern is that stuff usually desires to be more complex, to be baroque. The engineer can hear stuff asking for three more buttons, or for a little jack on the side, or for a web browser to be included. The real problem is that, by and large, people desire the simple. They don't want more complexity. And so the battle is joined on the workbench. The engineer must struggle to say no.

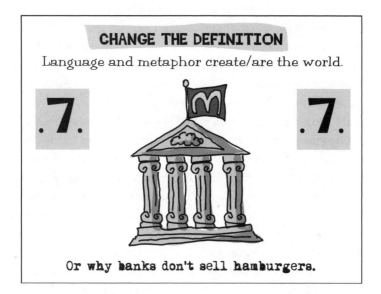

CHANGE THE DEFINITION

Language and metaphor create/are the world.

.7. .7.

Or why banks don't sell hamburgers.

The seventh pattern, Change the Definition, is the one that designers often invoke when they work on corporate identity construction. There is something fundamentally different between Xerox The Copier Company and Xerox The Document Company. It is more than a name change. These changes allow their respective companies to innovate in new areas, to develop and create in new ways, to behave differently. In a very real sense it changes the frame of imaginable invention.

But it has an effect on products, even existing products, and how we treat and use them. To call a cigarette a "nicotine delivery system" changes how we think about it, how we use it, how we advertise it, how much we will pay for it, and what legal structures we put around it.

The definition of a product (the product *genre*) determines to a large degree what it can and cannot be. It sets up the frame of expectations, not just for the customer but for the producer as well. Why, for instance, don't banks sell hamburgers? They sell all sorts of things, but why not hamburgers? This might seem like an odd question, but consider this: Seven-Elevens, and other convenience stores, sell money. They have little ATM machines sitting in the corner spitting out twenties at a $1.50 per hit. Not a bad business.

I once gave this talk to a group of bankers. After I asked this question one of them raised his hand and said that he had just gone into a deal with Starbucks and that they would be selling coffee and sandwiches at certain of their locations. They had changed their definition of bank to a place of life enhancement. Language alters the object and its trajectory. It was Walter Gropius, architect and founder of the Bauhaus, who called houses "machines for living" and in doing so opened up new vistas as to what you could design, or create, or innovate and still be making a house.

When I was at Mattel a group of us began calling dolls "physical fantasy tokens," meaning that they were little representations in the physical world of the complex goings on of a girl's inner mental world. They helped to stabilize and realize the girl's imaginative play. Years later, at PARC, we began to look into devices to help people remember things. To help them reify their memories in the real world. They were called "physical memory tokens." I called them dolls.

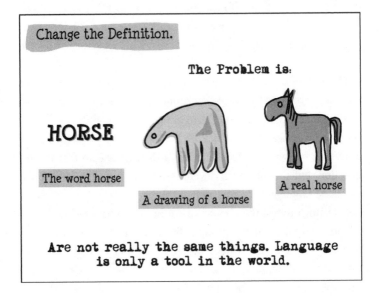

There is a problem with all of this, of course: The real world and the world of language do not match up one-to-one. I would say that language is not a representation of the world, it is a tool in the world. Unfortunately (or fortunately, since I think it makes art possible) we have tiny little brains that can barely tell the difference between the word *horse*, a drawing of a horse, a photograph of a horse, and an actual horse. It is why we cry at the movies, or even more amazingly, laugh at the little black marks in a book. The advertising world attempts to change the definition of a cigarette, for instance, by placing it in the proximity of a cowboy and his horse. The cigarette becomes the great outdoors and even freedom itself. The problem is, it isn't. So, if you are the one to choose the new definition, choose wisely.

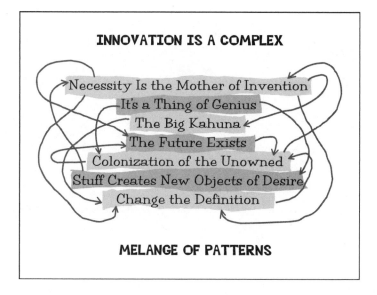

So those are the seven patterns. Certainly others come to mind, but seven is a good number. I want to reiterate that the seven work together. It's not like you can just use one; they blend and interact in a complex mélange. What changes between one creative profession and another is not the patterns, but the proportions of their use. Together, they are fabulous at making new stuff and the result is the Plenitude around us.

IV. THE PLENITUDE

I have had a life I'm not allowed to complain about. I've made lots of different kinds of stuff. Some successful, some not. Some really cool, some, well, let's just say, prosaic. It has been interesting. Far from complaining about it, I have begun to speak about it, going to companies and conferences giving talks on "innovation" and "creativity," which are just ways of saying "let's make more stuff." And that brings me, finally, to "The Plenitude" itself.

Creative and innovative are not words that I would have chosen myself to describe my life, but the culture has settled in on them. They seem like good things, fundamental to the running of the culture and to the success of the planet. But what is this thing called creativity? What does is it mean? At its core, I maintain, it simply means making stuff that has never been made before, that nobody has even thought of before and is not a warmed over replica of something already made. Now clearly I don't mean that the stuff has to be physical. It could be an idea, or a concept, or a string of words. But a lot of stuff is physical stuff, particularly in our world. Physical objects in our culture are like words in other cultures.

And just as clearly there are hierarchies of creativity, both personally and culturally. We often say something is more creative than something else, or that this thing is more innovative than that thing. That is, it is farther away from something that already exists—it is less of a copy. To be not creative is to "think inside the box"—to think of stuff that has already exists. In our culture of creativity we want the new, the different, the revolutionary. The boxes on the supermarket shelves visually squeal NEW NEW NEW.

But there is another meaning of the word *creative* that also has a qualitative connotation: It's not just something that has never been before, but it is something good, or useful, or communicative, or impressive, or beautiful and that a few people would buy for large amounts of money or that lots of people would buy for a small amount. Creative is a child when he or she draws a picture with purple crayon and, in a different (though related) sense of creative, when a scientist creates an unexpected equation or an artist produces a new and wondrous work. From the child's drawing to the painter's canvas is a continuum that it allows entirely new forms, new genres, of stuff to come into existence.

There are cultures where telling stories means retelling the same story that your parents told you. The power of the story, in fact, comes from the retelling of it over and over again. In its consistency, its sameness, it provides the eternal. In our culture, this is called copyright infringement and you can be fined or even sent to jail. Each story must be new and different. In some cases you cannot even reference an existing story (just try using *The Lion King* in a movie you are making and see how unbroken the circle of life can be). The sense of eternalness in our culture comes from everything being ever new. This is at the core of our culture. We must make things to get money to buy other things, including food and shelter. And since we can't make what others are making—by law and by the laws of the marketplace—it is only through creativity and innovation that we survive.

I realize that you might be thinking, "My God, what a terrible fate— I am condemned for the rest of my life to this [in my more cynical moments, I'd say] Junk Tribe!" But it's not like that at all. Personally, I love making stuff. I enjoy what others have called the creative act—when suddenly there is something where there was nothing a minute before. It gives me great and deep pleasure. In that sense I am a perfect member of this Stuff Tribe. Maybe I was born this way and just lucky to have found myself living in North America in the late twentieth and early twenty-first century, and not, say, in medieval France.

Perhaps the Tribe is just extremely good at forming perfect members for itself from the babies born into it. That's why we spend so much of our money on teaching our children to be creative: to make happy members of the Tribe.

A third possibility—that this is the perfect Tribe and that all other means of organizing humans are not just inefficient but immoral—is, of course, also interesting to contemplate. I suspect that many people do believe this: My father once expressed it this way: If the people of Russia just saw all the wonderful, creative stuff in our country they would immediately overthrow Communism and become just like us. Turned out to be a little more difficult than that, but he wasn't completely wrong either.

I have been able to make my living inventing new stuff, getting paid for it, and liking it. Does that mean that if I had been born into a culture that didn't value rapid and continuous change, I would have been unhappy? Or am I just an adaptable guy and would have made it work out no matter what culture I landed in? Hard to say from this position, but at the moment I'll stick with believing that I was lucky.

It is mostly incomprehensible to us what a small percentage of humans have ever lived in the kind of heaven we inhabit. Despite all the Plenitude's drawbacks, it is a kind of hubris for us to forget that fact.

Let me talk now about an epiphany I had. It revolves around this problem: Most people think making new stuff is hard, which is why they pay me to talk about creativity. And yet there is all this stuff. One day I was listening to the radio...

. . . and I heard a Wiccan on the radio. Wiccans are Good Witches. They believe, to way oversimplify it, that nature is the ultimate authority. Wiccans pray to the spirits of the biomass. When asked about it, this Wiccan said that abortion was not a crime against nature because, "nature (the goddess) is Plenitude."

And we do observe that nature produces, reproduces, and creates the new at prodigious rates. It replicates, procreates, and fills the space with innovations through evolution and with copies through sex. Fish don't just lay one egg, they lay millions. Given ten thousand years, even the comparatively infertile humans can pack a planet. Nature is thick and dense. It is also filled with death. Everything dies. Everything alive eats other things that are or were alive. Pruning is how you make a garden grow even bigger.

Life is fecund.

Nature is fertile and fecund. She grows and grows and grows and is almost impossible to hurt. Nature moves and grooves between plants and animals and viruses and retroviruses that are always dying, always being born, always in transition, always innovating, always exploding. Nature makes the atmosphere, the soil, and so many species that humans, though we try, can't count 'em. Nature is, in a word, Plenitude, and it is the Plenitude, and not the individual, that is precious. Any individual's death is the beginning of life for a thousand other parts of the Plenitude. Each new individual is just an experiment in even more life. Each new life is like a design exercise. Billions of times a second. Just as every cell in your body will be replaced hundreds or thousands of times, and yet are still you, the parts of the Plenitude come and go. If the cycle stops, if it stops changing, then, and only then, is the Plenitude in trouble. Eat away. It doesn't matter.

Now, contrast this concept, this idea, this feeling with the dominant, current metaphor of Western thought: Life is the hard-ass fight for survival and only the toughest, the meanest, the fittest, the most ruthless survive. Life is a giant sieve and only the biggest and the best make it. Life is hard.

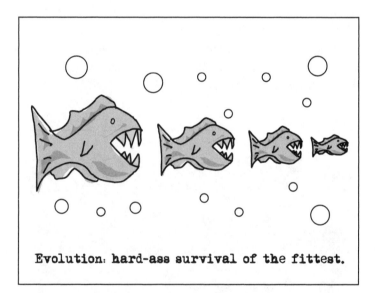

Evolution: hard-ass survival of the fittest.

And because life is hard, each individual life is precious. Is unique. Is fragile. Each organism, if it dies, is an indescribable loss. Life is fragile and living is tenuous and deadly. Living serves up a mean brew of justice and pitilessness. It takes half the scientists in Harvard to figure out why someone might want to commit an act of kindness.

That's how hard life is. If you are still alive, you are just lucky. Darwinian evolution, with its concentration on the individual, makes it seem like "almost nothing works" except that lone, lucky, fittest one.

But there is another way of looking at Nature. When my wife and I step out into our backyard we are faced with a circus of wildly differing plants and animals. There are big animals, like me and Marina; there are smaller animals like the cats and dogs, mice and raccoons; there are even smaller animals by the truckloads like worms and ants and caterpillars, bugs on six legs and centipedes on a hundred. There are birds that fly around in the air (and not just one kind, but lots of different kinds). And there are microbes and viruses and spores and amoebas and who knows what else. And that's only the life that moves!

Plenitude Evolution: "Almost anything works."

There are the big trees with their huge leaves. There are bushes. There are flowers in a riot of shapes and sizes. There are grasses that tile the lawn. There are mushrooms and fungi and lichens and ferns and bamboos. This is just my backyard I am describing, not even a real jungle. From it I conclude that ALMOST ANYTHING WORKS!

You name a mode of locomotion and some animal or plants uses it. You name a size and there is some animal or plant of that size. You name a color, a method of eating, a way of reproducing, a means of playing and there is some animal or plant who has developed that method. And new methods are being created all the time. Life is so easy that it is like falling off a log. It is explosive and creative and nearly infinite. Don't concentrate on the individual. The individual will get eaten, will get a disease, will die of something or the other. Concentrate on life itself. On Mother Nature. On the Plenitude! And in the Plentitude almost anything works! In the Plenitude innovation is all around. It is easy and simple and massive. Life explodes forth at every seam. For Marina and I, the difficult act in our backyard is not to create life, it is to keep the Plentitude under some sort of control. It wants to explode in every direction. Every Sunday I mow the lawn. Marina has taken to pruning every day. It hardly matters. The Plenitude thrives despite, or rather because of, the pruning.

But the Plenitude of the Mall shows that almost anything works.

But I don't want to talk about nature (or abortion with its complex moral issues). I want to talk about the making of the kinds of stuff that most people in the United States are employed to make . . . or dish up or dust or guard or store or display or transport.

There are many similarities between life and the stuff we make. Both seem to evolve. This year's product is only slightly different from last year's product. Lined up end to end, Chrysler cars have clearly evolved from 1911 to now with barely a surprising break from one year to the next. It is also interesting to note how, like their bio-counterparts, stuff falls into species and phyla. Into genres. There are vehicles. There are cars. There are sports cars. There's food; there's breakfast cereal; there's kid's breakfast cereal. And so on. And over time the number of these ontological distinctions increases. There are more and more kinds of communication devices and there are more and more kinds of phones and there are more and more kinds of phone services.

There are differences between the bio-Plenitude and the stuff-Plenitude. As far as I know there is nothing quite like brands in nature. But the big difference is this: Whereas in nature failures play no role in the next generation of plant and animal, in the world of stuff they play a large role. If a green doll is introduced one year and does poorly, new green dolls will not be introduced the next year.

It is commonly believed that new stuff is hard to make. That it is hard to invent, hard to manufacture, hard to market; that it is hard to distribute, hard to integrate into our culture. There is generally assumed to be such a cut-throat, winner-take-all economy that only the fittest, the toughest, the most ruthless products make it to store shelves, let alone survive. In this view, the evolution of new Product is very much like the evolution of life as Darwin saw it. That given how hard it is to imagine something that doesn't already exist, and given the number of constraints that a new thing must satisfy, it's a miracle we have any products at all in our homes. And it is certainly true that giant corporations feel as though they are about to be crushed in the international war for product niches.

But at the local mall, just like in my backyard, almost anything and everything seems to work! Not only are there fifty kinds of clothes (shoes, socks, shirts, pants) and new ones being added each day, but there are thousands of variations of each. The shelves of the markets are laden with hundreds of kinds of food and thousands of variations of each kind. And new categories of stuff are added all the time. Twenty years ago there were no cell phones. Now there are cell phone stores selling hundreds of varieties of cell phones. There are cell phones mixed with other electronic devices. New store categories open vast new niches for new forms of product. How many more kinds of underwear are there in Victoria's Secret than there already were in Macy's? The number of books in a Borders dwarfs the number of books in the older (now sadly displaced) bookstores. And Amazon dwarfs that number.

There are downtowns, and revitalized waterfronts. There are endless strip malls with a panoply of specialty shops. There's TV shopping, there's the internet with its near infinite snowstorm of different products; farmer's markets as well as supermarkets, as well as gourmet markets, as well as WebVans. Each of these product ecosystems is loaded, stuffed, teeming with different forms of products. Not only are there dozens of kinds of toothbrushes available, to pick one example, but they come in multiple designs, colors, stiffnesses, and packaging. And these toothbrushes coexist with other members of their species: Waterpiks, rotating toothbrushes, laser cleaning, vibrating brushes, and on and on.

The mall is a dense jungle of every kind of product and species of product. Sure, lots of the products will die out this year. More will be back next year, slightly improved, slightly innovated. Not only does this not matter, but the whole economy would grind to a halt if it wasn't mostly different next year. Toy companies replace 80 percent, or more, of their SKUs each season. If they didn't, they would go out of business because kids already have last year's stuff. More important, what they really want is, simply, the new. It is only the Plenitude as a whole that is stable. And what you see that has been produced, has been shipped, and is now stocked on the shelves is just a tiny percentage of all the stuff that was dreamed up, brought to prototype, maybe even market tested. For every one toy that made it out of Mattel, there were a thousand that ended up on the workshop floor. The magnitude of the Plenitude is difficult to grasp. It simply doesn't matter that most things don't work out, for even as it is, the Plentitude may be too fecund for the planet to absorb.

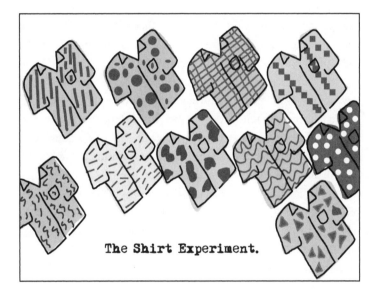

The Shirt Experiment.

Here is an experiment that I often give during my talks. Look around the room. You will notice that everyone in the room is wearing a different shirt! Different colors, different patterns, different cuts. Imagine, for a minute, the vast design energy that is required to make this many different shirts. The cloth designers, the fashion designers, the factories, the seamstresses, the shippers, the clothing stores, the advertisements. Really tens to hundreds of people were involved in the making of each one of those shirts. And here's the even more amazing thing: If everybody comes back tomorrow to the same room, say to hear another speaker, then everyone (except a few computer nerds) will be wearing yet a different shirt!

(There was an occasion when this experiment failed. I was giving a talk at the U.S. Coast Guard headquarters. The men and women in the audience stared at me as they sat in their identical powder blue short sleeved shirts. Of course that's why we call them uniforms.)

At Xerox PARC there was a researcher who came from a small country in Africa. When somebody in his village wanted to get a new shirt, he said, that person went to the shirt guy in the village. It would take three or four days and when it was done the unique shirt reflected the person who wore it at a deep level. It was beautiful and highly meaningful to the wearer, who would wear it every day.

I can almost guarantee that almost nobody in my audience cares that deeply about the shirt that they are wearing. They may not even remember where they bought it. If they close their eyes, they might not even be able to recall what shirt they are wearing. The important thing is that they are wearing a physical representation of the Plenitude.

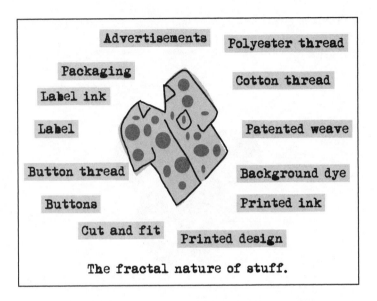

Advertisements　Polyester thread

Packaging

Label ink　Cotton thread

Label　Patented weave

Button thread　Background dye

Buttons　Printed ink

Cut and fit　Printed design

The fractal nature of stuff.

I once asked Hal Varian, the economist, how much stuff is there is one room? "Well, what's the definition of a piece of stuff?," he asked. "How about: something that was individually designed, shipped, marketed, and sold," I replied. "OK," he said, "then there is no answer, for stuff is fractal." What he meant (I believe) is that while we might say that a shirt is a single piece of stuff in the Plenitude, I could also say that each button was individually designed, shipped, marketed, and sold. The same for the thread. And the dye that colored the thread. The weave of the thread might be patented and based on equipment that was designed, shipped, marketed, and sold. The cut is no doubt itself a copyrighted legal entity. The advertising for the shirt, which made me buy it and which covers the shirt with a sparkling aura, was also designed, shipped, marketed, and sold. Not to mention the genetically engineered cotton. The Plenitude is fractal and it goes all the way down.

We must make new things by law.
Its illegal to tell the same story again.

As I have already noted, in many cultures, including ancient cultures, the telling of stories is about telling the stories that your parents told you. And the stories that they told were the stories that their parents told them. We still see this in religious settings—the telling of Bible stories, or the reading of the Talmud, the Koran, the analects of Confucius or the I Ching. We even see it a little in the origin myths of America itself. The Constitution is often replicated and referred to.

But in the Plenitude you are not allowed to tell the same story that you heard, or even use the same characters in a new story! Not only is it considered in poor taste and probably not profitable to produce something identical to something else, it is against the law. Variation is built into the legal system of the culture and lies at the heart of the Plenitude. At this point it's a reflex for us to seek out the new, the different, the creative, the innovative. That's what we like and that's what we buy.

The Plenitude has related names such as "Progress" and "Industry."

Another key concept in the Plenitude is progress. The primary tenet of progress is that every year the things we make are better than the things that came out the year before. A car built in 2001 is better not only than a car built in 1910 but, theoretically, even one built in 2000. Occasionally we find some object from the 1950s that's better than what we currently have and we note it with a sigh.

Not only must this year's stuff be different from last year's stuff, but flat production is not good enough; there must be precisely 3.5 percent more stuff this year than last year or it is considered a recession! This places quite a burden on the consumer (that hypothetical animal whose desires and fears are quantified in something called the consumer confidence index). To buy more stuff people have to work more hours and if they work more hours they don't have enough time to buy and use more stuff. The real point of efficiency is to allow people to make more stuff in a smaller amount of time so they have enough time left over to purchase stuff.

Industry is another term deeply intertwined with the Plenitude. While we might imagine a new world of e-everything and service-everything else; in reality, we mostly produce tangible stuff and move it around. The web is a big brochure. The Plentitude is made, and most of it is made by industry. The structure of industry (for instance, how it separates marketing from engineering) and the technologies of industry (for instance, the assembly line or the stock market) all feed directly into the Plenitude. (I am of course willing to entertain alternatives—for instance, there have been attempts at cultures with industry and progress but not Plenitude: The Soviet Union wanted industry without the Plenitude. Certain fundamentalist countries will try their own variations. But progress, industry, and the Plenitude may just be too deeply intertwined to survive independently.)

Also deeply and perhaps inextricably intertwined with the Plenitude is the corporation. There's little doubt that the two grew up and achieved a kind of mutual maturity together. The corporation is a peculiar entity with, on one hand, limited liability (stock owners do not have to pay for losses) and on the other, many of the legal rights of individuals, including freedom of speech. Corporations have a strong structural desire to grow and are only barely tied to place. They feed off the Plenitude and most are designed to their very core to create more of it.

Most corporations have some generative organ for creating new stuff. This might be an R&D department, an engineering group, a marketing division, a group that buys outside inventions, whatever. Corporations have to sell something after all, and most have to sell new stuff. Even corporations that rely on fairly unchanging commodities (such as corn flakes) have methods of making the flakes new and improved each year. Or at least their boxes.

How is the Plenitude related to mass production? Mass production, of course, is a process where one designed item—the item in which the creativity, or innovation, or ingenuity resides—is duplicated thousands, if not millions, of times. Nobody in a given room may be wearing the same shirt, but there are thousands or millions of people wearing that same shirt somewhere else. In general we think of two broad kinds of mass production. The first takes place in factories. The invented object is turned into a set of tools, dies, and procedures. The factory is turned on. Raw material is dumped in one end and the tools, dies, and procedures turn the raw material into duplicates of the invented object, which leave the other end of the factory to be distributed to you. There are factories all over the world doing just this, though more are found in poor countries than in rich ones. The second kind of mass production requires a certain kind of object, usually called a *medium*, that can take another kind of thing, usually called the *content*, and display it. Media would include TV, radio, and the internet. Content would include TV shows, movies, and news stories. Using various forms of broadcast the content is sent out to the millions of mediums that can reproduce the content right there in the living room. When you watch "Survivor" on TV, millions of other people are seeing it too, on their own little content factories.

There are other kinds of mass production. Paintings in museums, which seem to be one-of-a kind things, are usually remediated for postcards, books, posters and the like. In many cases it is this remediation that makes the painting valuable. The image, removed from the canvas, is what is mass-produced.

We could imagine a culture that has mass production but not the diversity of the Plenitude. It is harder to imagine the Plenitude without mass production; certain economies of scale facilitated by advertising simply wouldn't work.

There is also one important change occurring. It goes by the catch-all name of *customization*. Factories are being adapted to be able to spew out product that meets the desires of different niche markets; there are many websites where the content dished up to you is based on your preferences and other factors they have been able to glean. That is, the content is based on your profile. (Better hope your profile is right.)

The laws of the Plenitude extend even to ideas. There is no information without representation.

Less straightforwardly, the Plenitude also exists in that ethereal realm of ideas, often for many of the same reasons it exists on the physical plane. In a bookstore, the sheer quantity of ideas, of thoughts, of worked-out theories, of constructed fictions—most of which I will not have the time to know about, let alone read—is startling. The Plenitude one finds inside of a bookstore makes it probably the second most frightening place I can go in a mall. The most frightening, in case you're wondering, is the toy store. The toy store's purpose is quite precise and highly targeted: to incorporate our young into the Plenitude.

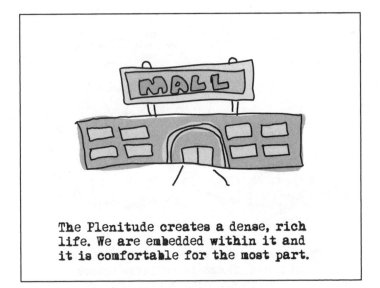

The Plenitude creates a dense, rich life. We are embedded within it and it is comfortable for the most part.

From experience, I know that the readers of this work have already divided into two groups. One group is thinking, "Wow, what a great culture to live in! I gotta get off my behind and start a business! This is great! I need to start making things for the Plenitude!" The other group is thinking. "Oh my god! This is terrible! What an awful, superficial way to live ! I am completely swaddled and surrounded by the Plenitude! I'm a victim! I have to figure out a way to get out, get off, get away!"

It's not just the readers who are divided; I am too. But then I remember that there is no place to stand without contradiction, either inside or outside the Plenitude. Contradiction is simply part of our condition. Many postmodern critics refer to what I am calling the Plenitude as Late Capitalism . . . as if they know that global corporate consumer capitalism has pretty much run its course. I don't see it that way. I think we are in the Early Plenitude, like living in Rome in 300 BC. If history is any guide for how long these sorts of waves last, we have another five hundred or so years to go. The rules aren't fully formed yet; there's still wiggle room.

You're thinking, he's getting pretty cynical here. There seems to be an undercurrent of negativism creeping into the book, which started out as a kind of pleasant, if a little self-centered, retrospective of a creative life. Does he, or does he not, like the Plenitude? If only it were that easy.

How remarkable, really, that we can be so at odds with major parts of our own culture. I'm not sure what it means, really. It's as if our moral framework has worked itself loose from the very matrix that produced it. What other rules can we use but our own? Are we a conquered country dictated to by some foreign ruler? No, we are the empire! Have there been other civilizations that produce things, not just peripheral things but real engine of survival things, that the populace itself has trouble with? That they won't let their own children look at?

I knew designers at Mattel who would not let their children play with the toys that Mattel produced. There are toys that I won't let my son Henry play with. These aren't toys from some other culture. People just like me—living in L.A. or Rochester, Tempe, Urbana-Champaign, New York, or Raleigh-Durham—designed these toys. I have a friend who wrote TV shows but did not let his kids watch TV. I won't let Henry watch some TV shows, shows designed exactly and precisely for him by people who look just like me. As a parent what am I to do with cartoons that present themselves as outsider and antisocial, and that just happened to be produced by major corporations?

You hear complaints that a kid sitting alone in his room, facing a computer for hours at an end, shooting at virtual but very realistic humans with a wide variety of virtual but very realistic weapons, is antisocial behavior! As if the video game and the computer, and the chair and the room, for that matter, weren't all created by our society. The people who created this experience are still alive and live next door. From a Plenitude viewpoint, this behavior is more social than going to church and praying to a Middle Eastern God from several thousand years ago.

Let me be clear.

There is no activity I can think of that I enjoy more than making stuff for the Stuff Culture, junk for the Junk Tribe. I enjoy going to Borders, or to the mall, or to the craft fair that takes over Palo Alto's main street once a year. I enjoy being overwhelmed by the vast diversity and the breadth of our culture's creative energies. This is what we do and we are damn good at it, thank you very much.

But while one can spend one's week-days constructing the ever-changing, the ever-renewing, ever-growing-at-3.5-percent-per-year, our religious institutions spend Saturday and Sunday mornings arguing against this worship of Mammon. Against material desire. Against the transient. Against the accumulation of stuff for its own sake. In school, our children learn ecology as a basic moral tenet.

Let's face it, when you have a culture built more or less around the centrality of corporations who promote consumption while many people worship a god of poverty and meekness and the children are taught about saving nature in school, something is askew.

As many great minds have reminded us, there is no place to stand without contradiction. Maybe this slight queasiness is just what happens when you spend so much time looking at yourself. You get dizzy. After a while you begin to wonder why you are alive, what is the nature of life, what is reality, what does it mean to exist? These are the questions that make all cultures dizzy.

It is surprising to students that their job in life is to make more junk for the Junk Tribe. It is even a little alarming!

When I give talks to students on this topic they are often somewhat surprised. It really hadn't occurred to many of them that what they were learning to do in university is to make more stuff. They thought they were going to cure cancer, or help people communicate better; to make children happy, or solve security problems in email servers, or design shirts that will make the people who wear them look really cool. All good and noble pursuits. It didn't dawn on them that they are part of a process whose primary function is to churn out stuff and it will be impossible for them to earn a living without making more stuff. We simply do not know how to make a living otherwise. Just what were these students thinking?

Let me refer back to my 2-by-2 of the four creative professions and construct a matrix of possible thoughts.

The student engineers saw the world as a series of problems to solve and assumed that by solving these problems the world would become simpler, nicer, more humane. The computer, for instance, would simplify office work by making it more pleasant to spend eight hours at a desk. In reality, of course, such solutions are not only almost always more complex than the system they are trying to fix, but they are additive, layering on top of the old solutions. That's one of the properties of stuff I haven't mentioned: It piles up. Offices now have both computers and filing systems. It's just a fact.

The student scientists thought that they were simplifying the world. After all, each round of physics theory seems simpler than the last; each equation is meant to be true forever. Science seems the opposite of junk. And yet, each new equation opens vast vistas of new stuff that then pours from R&D departments all around the world. Scientists have a wonderful phrase for this. They say "The genie is out of the bottle and you can't put him back in." And we know what that means: Once we have created the formula, the stuff will flow and you can't do anything about it.

Now student designers, when they are asked to design, say a new toy, imagine that the child is playing with just that one toy. All alone in a pristine child's room. But that isn't how children play. A toy is just a component in a vast configuration composed of multiple toys. No one toy satisfies a child. That's not what toys are about. Toys are about lots of toys and about ever different toys. I am looking for a new suffix—the continuous plural—to indicate this. Perhaps three "s" in a row. Only *toysss* will satisfy a child.

All designed stuff is additive. It bunches up under beds and on livingroom shelves. Simplicity, when we do see it in design magazines or interior decoration books, is always very, very expensive and, remarkably, takes a lot of time. Only the rich can afford that kind of simplicity.

Student artists, of course, rarely think of themselves as training to make more stuff to fill an already stuffed world. But consider this: In some states, including California, it is illegal to destroy an artwork. In the middle of World War II people actually risked their lives to save art. We're always looking for a definition of art; well maybe this is it: it's stuff you are not allowed to throw away. It collects in vast and dusty warehouses, mansions and museums. A city may have homeless people, but it almost always has museums, homes for art.

On one hand there is the simple shock of recognition. My God, that's what I will be doing for the next sixty years! But on the other hand, there seem to be real problems with the Junk Tribe, with the Plenitude, within our own tribal, moral frames. Let me lay out some of these troubles.

So why the shock? Well, one reason is my use of slightly pejorative word *stuff*. The desire of most creative folks in America is to make something revolutionary. Part of our heritage, I suppose. A new car, or email browser, a new sneaker or cancer drug or chemical for binding IC chips to circuit boards, it just doesn't sound paradigm-shifting enough when it is referred to as simply "stuff" or worse yet "junk"!

But there is another reason. It's true that we don't think of ourselves as a particularly moral culture. We say that we are a nation of laws, not morals. Our concept of the free market is usually about as deep as we go. We even believe that each person can choose his or her own religion with its own moral percepts. That's how little we think it matters. Just come up with your own religion and, as long as it doesn't violate our laws, then fine. That's why we allow Wiccans on the radio.

But in fact, we are a moral culture; all cultures are moral cultures. And I believe that the morals we have are at odds with the Plenitude. There is a sense in which living in our own tribe, the stuff tribe, is itself an immoral thing to do. When a parent says "no more TV" to a child it is this very tension coming to the surface. We seem to have a moral aversion to ourselves. It is the exposition of this moral/culture schism that shocks the students and makes them blanch. They are morally outraged to discover that they are being trained to help make the culture that they live in.

I would love to go on about the Plenitude in moral or even religious terms. It seems fitting for somebody who spent his childhood creating cults. But I don't feel qualified. It is, after all, I who, at some people's tribunal of the future, will be brought up on charges of crimes against humanity for making all this stuff. Instead, I will explore on a more concrete level some of the problems of the Plenitude. I will try to lay out some of these systemic problems so that later, maybe, we can find solutions to them. So that we can reduce the desires that are inherent in their formulations.

All complex systems have two competing properties. They are homeostatic and they are self-adjusting. The first maintains the identity of the system, essentially keeps it the same and invariant. The second changes the system in response to an altered environment, or possibly, for self-improvement, or sometimes, on occasion, just out of boredom. The most interesting systems are the highly homeostatic ones that can find problems within themselves and then self-adjust to maintain homeostasis. We like to think of the United States this way—with its amendable Constitution and checks and balances. In the same way, most solutions to the problems of the Plenitude address sustainability.

So what exactly are the problems with the Plenitude?

One: The Plenitude creates a world somewhere between the bland and the ugly.

The first problem is that the Plentitude creates a world that any dispassionate observer would have to say lies somewhere between bland and ugly. Sometimes it is the individual objects themselves; but really it's more the totality of the vision, the whole, that is so deadening. One of the current trends in museums is to take some specific object from the Plenitude, pull it out of the clutter, and present it alone, showing how, in its essence, it is beautiful. But of course this only proves the point. Some individual threads may be gold but the conglomerated Plenitude is aesthetic burlap at best. There are moments, of course. Los Angeles, for instance, is the epitome of ugliness, despite moments of local charm. Yet at night, up in the hills looking down on the spider work of street and car lights, there is a gorgeousness that rivals the stars themselves. But these shiver moments are rare. In general, I would have to say that the Plenitude edges toward the ugly. There is something physically damaging, something punishing, about living inside of ugliness. Jails aren't ugly just because they are cheaper to build that way, the ugliness is part of the punishment. The Plenitude's aesthetic takes its toll on humans.

Guy Debord, the Situationist philosopher, called our culture the Society of the Spectacle. It is a deadening spectacle, one that strips the senses and the sensibilities of meaning. When I look at an audience of two hundred people, yes, I see two hundred different shirts. That should bring tears to my eyes, I should be blinded by the effort of art and design to enrich and enliven my world. But no tears come. Not even close. Not one shirt in the crowd, my guess, actually means anything to the person who is wearing it. The totality of the spectacle does not provide a deeper meaning, a more complex woven pattern, does not point to a deeper truth. It points nowhere except maybe to a Gap (pun intended).

The Plenitude's "diversity" is overwhelming and it drowns out beauty, drowns out anybody trying to say anything. The collective dazzle results in a dullness that can actually ache. If, by some miraculous force of will, somebody really does make a shirt with meaning, the Plenitude quickly, and effortlessly, mass produces that shirt, wiping out the meaning. Remember that hippie anarchist moment of craft called tie-dyeing? Well, now they're printed and you get them at Target.

Real Dog

Toy Dog

TV Dog

Virtual Dog

Two: The Plenitude blurs the distinction between the real and the faux.

Problem number two. We have remarkably small brains. That's why pornography works. Our tiny brains make little distinction between a real naked person, a picture of a naked person, or even the description, in text, of a naked person. Yes, of course, we know that these are different, but on the basic level where we react and act, we can't actually make the distinction. That's how puny our brains are.

A real dog, the video image of a dog, a toy dog, and a virtual-reality interactive dog, four PowerPoint dogs, or a page printed with four dog images. Whatever. The amazing thing is that you see dogs at all. There are no dogs in front of you. That's how small your brain is. Until very, very recently, far too recently for evolution to have adjusted for it, there were no images and certainly no verbal or textual symbols. In some odd way symbols work because our brains are so old-fashioned that we can't really make the distinction. We make fun of early filmgoers who ducked when the black and white locomotive came at them. But we can still get scared at horror movies, cry at tragic love stories, and get aroused when looking at pornography.

The Plenitude not only revels in the smallness of our brains; it actively courts the muddling of the real and the image of the real. We in the Plenitude live in a perpetual fog of blurring between the symbolic and the real, the picture, and the thing. When I listen to a CD it is as if I am listening to real people playing real music. When I read a book it is, first of all, as if I am listening to a real storyteller; then if the storyteller is any good, it is as if I am living in the world he or she spins. From the soda I drink to the car I drive, from the patterns on my shirt to the nature shows on TV, every reality is layered with one, two, three, ten layers of unreality. Ten layers of indirection. Or as they say in the computer science world, of operator overloading. Our small brains reel.

Even things of such dense reality as war spread out as a system of signs and countersigns, symbols and words, TV shows and newspaper photos. This doubling, this *doppelganger*-ing, of the world, where everything has both a real and a referential part, where there is actually a genre called Reality TV, which is not only on TV but which is staged, should make our lives shimmer with possibility. But I find that there is something oppressive about it. A Plenitude of mirrors makes it hard to tell what's real, makes it hard to act and react meaningfully. And yet, there is a real world, a reality one can't get out of. It is not human-defined symbols all the way down. The Plenitude, with all of its stuff, pushes us from this reality, making it hard for us to see it, to understand it, to act, to react wisely within it. Hard to engage in reflective homeostasis and autocorrection.

I'm not sure that we can create a Plenitude that is not based, in some large part, on reference; We have certainly not done so. Our Plenitude is itself a language. A car is not only a car, but a means of saying something to others, to ourselves; a complex language that not only deadens, but can kill. The same is true for the other 10,000-plus objects in each of our homes. It is quite a language.

One solution to the information explosion.

To restate problem number two: The Plentitude blurs the distinction between the symbolic and the real. This blurring in turn creates a world that overwhelms us, confuses us, and creates anxiety, paranoia, and confusion. Information overload is not about too much information. As Dr. Weiser pointed out, there is more information in a forest than in a suburban house. It's about *symbolic overloading*.

A flower in the woods exists and can have a meaning for someone. A flower image on the side of Kleenex box has ten meanings, all operating simultaneously and often in opposition to each other. Among other things, after all, it has to mean both *natural* and *manufactured* at the same time.

What if we had two heads, one for the presymbolic world and one for the symbolic? Given the rapid advances in genetic engineering, this may soon be possible. Wouldn't it reduce the information overload by half? I once presented this, clearly facetious, solution at an engineering conference. A hand immediately shot up in the audience. "This is a well-known problem," said the bearded man. "You need a third head to mediate between the other two, otherwise the system starts to thrash." Thrashing is when a computer system spends all of its time switching between programs and never doing any real work. In the Plentitude we all thrash all the time.

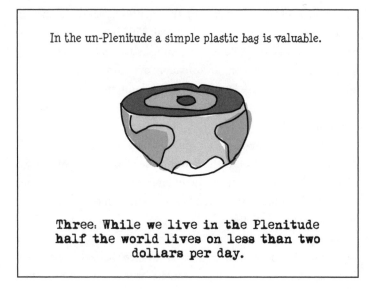

In the un-Plenitude a simple plastic bag is valuable.

Three: While we live in the Plenitude half the world lives on less than two dollars per day.

The third problem with the Plenitude can be simply stated as this: Half the world's population lives on less than $2.00 per day. And at least a billion people live on less than $1.00 a day. In other words while we live in the Plenitude, most do not. What I pay for a cup of coffee is what most live on (including food, housing, clothing, education—if any, entertainment—if any, and medicine—if any) per day.

Some of the moral groundwork that the Plenitude rests upon is a general belief that while yes, some people are richer and some our poorer, nobody is living on garbage heaps on the equivalent of $1.00 per day. But it is a fact, and it calls into question the wisdom of the whole enterprise. We aren't talking about some fringe element here, some bums who don't want to work. We are talking about more than half the world. Even those among us who are not offended by this reality on moral grounds, believing that it is simply not our problem, are beginning to realize (owing to certain world events) how small this shrinking world is. In fact the most powerful shrinking agent is the Plentitude itself, projecting images around the globe.

Does the Plenitude require this division between the vastly wealthy and the very poor? Is the only way to have the vast array of shirts, pants, cars, computers, breakfast cereals, and oil here to have extremely cheap labor somewhere else? If the rest of the world made our minimum wage would the Plenitude grind to a screeching halt? I don't know. Maybe.

One counterargument is that the only hope for the rest of the world is to embrace the Plentitude. There is reason to believe that this might be coming true. A hundred years ago, nine tenths of the world was in dire poverty, so things are improving.

Is this the bargain then? Embrace McDonalds and in return you get the rest of the Plentitude including medicine, housing, and television? One question is: Is it worth it? Many in the world today reject this bargain, for it means the destruction of their own culture, which is often deep and beautiful. The second question is: Can we keep the bargain? Is it a real bargain, or just something we say to keep the World Bank and the IMF in business? After we put a McDonalds next to your holy sites, after we cut down your rain forests, will we still hire your people at one dollar per day? There is rioting in the streets over these questions. The answers are unknowable.

Four: The Plenitude may very well destroy the world.

The fourth problem with the Plenitude is not an aesthetic, a linguistic, or even a moral question one about whether it is OK to eat hamburgers while others starve. It is a rather practical question about whether the Plenitude will destroy the world. Of course all such questions are moral ones. Only on a moral basis can we say yes, it is wrong to destroy the world. The universe certainly doesn't care. The universe will sooner or later destroy the earth on its own.

The Plenitude could destroy our world by simply using up all the world's resources so that, like a fish out of water, we simply suffocate. We could use up not just the oil and trees, but the arable land, the metals, flora and fauna, the oxygen. Once these resources are gone they are gone for other systems as well. In the end, perhaps no human society will be able to exist, for little life may be able to survive the toxicity. Poisoning the world is probably a better way of putting it. We have already altered the air, the water, and the soil and changed not only the chemical nature of the biosphere but its weather patterns.

We have become so worried about this little side effect of the Plenitude that we teach the evils of pollution as part of civics in schools paid for by sales taxes on the Plenitude itself.

Finally, the Plenitude has a peculiar relationship to a much older institution—war. The Plenitude certainly did not invent war, but it is extremely good at producing it in a magnitude previously unheard of. There was nothing like World War II, a true war of the Plenitude, ever before in the history of humankind. The nuclear bomb is almost a perfect weapon of the Plenitude, and we have manufactured enough of them to eliminate all life on the planet for good.

There have never been armies like the armies from the lands of the Plenitude. Some large percentage of the stuff of the Plenitude is weapon stuff. We certainly know that the only countries with real working weapons of mass destruction are those of the Plenitude. We certainly see those countries get mad when any other countries look like they might want to develop some. This substructure of armor seems necessary for the Plenitude's existence. But it is an unstable foundation in a world that has also become digitally wired together.

It has been argued that the Plenitude loves peace, for in peace there is the real money to be made. Better to sell refrigerators than drop bombs. But if most of the raw material, and most of the raw labor, comes not from the lands of the Plenitude but from the lands of the two-dollar-a-day worker, then the military and its might are more than simply a habit that has hung around too long. They are the underpinning.

Five. How many genetically modified organisms gone wrong will it take to bring our life support system down?

The Plenitude seems to be able to produce a plenitude of means for destroying the earth just as it is capable of producing a plenitude of different colored shirts. Who could have imagined a hundred years ago, fifty years ago, that the Plenitude could create entirely new creatures, big and small, whose habits and proclivities are as yet completely unknown; or that we would nonetheless release them into the biosphere in the name of feeding poor people? We can't engineer a car that doesn't kill people, we can't create a computer that doesn't crash, and yet we are now re-engineering the biosphere with new life forms.

The fifth problem with the Plenitude is that it has no way of stopping. There is no boundary it won't cross. After filling supermarkets with new and improved products, it has decided to re-engineer the biological Plenitude so that it could make even more products to fill the shelves. We are now creating new species, each one to solve some problem created by some other part of the Plenitude. Pesticides killing your crops? Let's bioengineer a solution to that! Let's genetically engineer an entirely new species of crops that aren't affected by the poison so we can pour on even more poison.

OK, Rich, why do you continue to make stuff for the Plenitude?

You could ask this because I just laid out five compelling reasons for us to seriously consider that the Plenitude is not a good thing.

- First, it causes an ugliness that damages.
- Second, it blurs the distinction between the real and the virtual to a point where we can't act intelligently.

- Third, it seems to keep one half of the world in dire poverty.
- Fourth, it looks as if it is on the verge of destroying the planet.
- Fifth, if Plenitude doesn't destroy the planet, it will re-engineer nature into a product that will.

Of course most people within the Plenitude see these problems only hazily, in the distance, in the periphery. The Plenitude has such a remarkable ability to provide all the basics in such wild abundance and endless variety and quantity that to those within it, it is barely distinguishable from heaven.

So what would the answer to this question look like? That nothing is more enjoyable to me than making stuff for the Plenitude? That I am part and parcel of this culture and it is impossible to act outside of it? That the Plenitude is an amazing engine and it will find the solutions to these and all other problems just at the last second? That life is always hard? That there is no place to stand without contradiction? There is no system of economics, of human interaction, that only has a positive side. Every culture has its own contradictions, its own confusions, its own ends-of-the-world. This is our culture and we cannot see outside of it. All the solutions that I will shortly present are solutions that somebody who grew up in the Plenitude came up with.

The Plenitude has built into it the concept of continual change. If I suggest that we try to figure out ways of changing the Plenitude so that it works better and solves some of the problems I have outlined, I won't be branded as a heretic to be burned at the stake.

To those inside the Plenitude the idea of changing the Plenitude is no more radical or unusual than if I stood up and said: "You know what, I think we need to change Microsoft Word to make it better." Or, "I think we need to change cars so that they run more efficiently." Or, "I think we should create a new fashion for shirts this year because last year's were so dull." There might be a difference in magnitude, in the level of hubris, in the scale, but changing things is what we do.

Changing the Plenitude is complicated. For starters, there is only one Plenitude. It is the operating system, and when you change the operating system lots of other stuff tends to break. We have to be careful, of course; this is exactly how Captain Kirk destroyed many an artificially intelligent computer system in deep space: by using its own logic against it. Be prepared for smoking ears. But I think we are compelled to try and change it, despite the recursive problems created when we bring to bear the rules of the system onto the system itself.

We can take the Seven Patterns of Innovation and apply them to our own culture, our own tribe, to the Plenitude itself. At least that's what I tell myself on certain days . . . you know those days. When I have my engineer's hat tightly on my head.

When I present these ideas in talks the first question I get, invariably, is: "Isn't there anything we could do about this? After all, Necessity is the Mother of Invention. Innovation is a Thing of Genius. All we have to do is Change the Definition. You said so yourself!"

"So are there any solutions?" the kids ask. "Necessity Is the Mother of Invention," I reply. Innovation churns at every level in the Junk Tribe, within the Plenitude.

Then I would field solutions that span quite a gamut. Some would keep the Plenitude intact, just cleaning up the edges. Others would transform it into something quite different. Some suggestions only a Dadaist could admire. But on the whole, each audience presented more or less the same solutions. I think that these are the solutions that come most readily to mind, or maybe they are the only solutions.

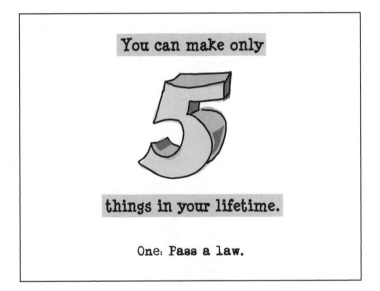

One. We could pass a law. Here's a good one: You can only make five new things in your lifetime. That's it. Five. Anything else gets destroyed and/or you go to jail. If you multiply five times the number of people on the planet, 6 billion, you still get an awful lot of new stuff for the Plenitude. It just limits it a little. And it does one other important thing. If you can only make five new things in your lifetime, perhaps you will think about each one much more carefully than people currently think about the new things they make. Quality would go up. Things would be better.

I like this one of course. But it does presuppose a legal system and the police muscle to back it up. It presupposes, in other words, a rather powerful State. Downer. It pretty much guarantees a very active court as people try to figure out what constitutes a new thing, versus a variation of an existing thing, or just a duplication. Is a variation on mom's soup a new thing? Is a weekly magazine one thing or fifty-two things? Is a rock album 12 songs or a single musical experience? How does manufacturing play into this? Broadcast? Computer programs that can generate a million other things? I suspect there would be a lot of 10,000-page books and 1,000-foot-long paintings.

Two: Reject the Plenitude.

Two. Simply reject the Plenitude. This is what many fundamentalists in different parts of the world suggest. It is what the American Puritans suggested, and several commentators on the Plenitude have suggested that our current concerns stem from these Puritanical moral underpinnings. Thoreau sitting around Walden Pond and Ted Kazinski sitting in his cabin in the deep woods. It is a deep tradition and it may be why, when I talk about the Plenitude so openly, we are all a little nervous. Because for each of us, to some degree or another, there is *some* part of the Plenitude that we reject. Perhaps you won't eat junk food. Perhaps you won't wear T-shirts with writing on the front. Perhaps you won't watch daytime TV. Perhaps you think that tearing down rain forests to make way for cattle farms is wrong.

When people find they hold two belief systems simultaneously, they often pick one and go overboard. It is hard to find the middle ground and defend it. Perhaps the Plenitude is so massive and overwhelming because we also have a fundamentalist streak that we have to constantly suppress. Reject or embrace. Nothing is harder for a parent than to say to their child "you can watch only one hour of TV a day." None at all, or just leaving it on all the time, is easier.

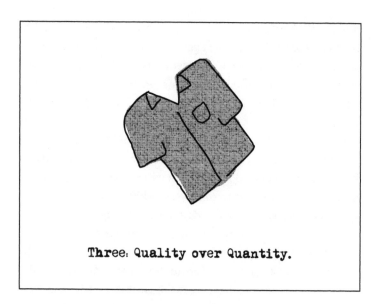

Three: Quality over Quantity.

Three. The third commonly suggested solution to the Plenitude centers around the opinion that the primary problem with the Plentitude is not its Plenitude-ness, but its Junkiness. No one seems to
mind a massive plethora of fine restaurants. It's only the plethora
of fast-food chain joints that is loathsome. The quality over quantity
argument suggests that if we could jimmy the Plenitude sidewise
just slightly, life would be good.

What would it take to thus re-engineer the Plenitude?

First, it would require a different mode of production, one that
honors craftsmanship, for we often see the human hand as part of
quality.

Second, it would require high-quality, lasting materials, put
together with intelligence instead of the cheap, the temporary, and
the robotic.

Third, it would require expensive and complex design methods including deep customer engagement and feedback. To put it bluntly, it would require that we replicate the production methods that the wealthy currently use. *Think about the difference between how a wealthy person has a new house designed and built and how a middle-class person finds a house to move into.*

When these three principles (and others) are applied across the board there will be at least two immediate implications. The first is that most people will have to have fewer things, though the things they have will be of higher quality. Instead of ten twenty-dollar shirts hanging in the closet there will be just a couple of hundred-dollar shirts. But they will be great shirts and you will look great in them. The second thing that will happen is that the hundred-dollar shirt will require five times as much labor to make as the twenty-dollar shirts. This isn't bad since otherwise people will be out of work from all the closed factories and malls.

It should be pointed out that this solution only works if quality can be substituted for quantity in such a way that the velocity of money, how fast it moves from one set of hands to other, remains the same. This is a stringent requirement and it is unclear whether it can be effected.

And it certainly will work only if one other, much more unlikely thing is true: that a huge re-education effort takes place to teach the average consumer about contemporary, high-end design making it not just desirable but necessary. Necessary in the sense that if somebody sees you with a twenty-dollar shirt they will think ill of you; while if they see you with a hundred-dollar shirt, even if you have worn the same shirt five days in a row, they will not.

I think this is an iffy proposition. There is something wonderfully democratic about the Plenitude; something really neat about ten cheap shirts vs. two expensive ones that will be hard to alter. And, as many have noted, these days, even rich kids like to eat at McDonalds.

Four: Zero-growth economies.

Four. Related, but not the same, is the popular concept of zero-growth economics. Here the problem of the Plenitude is reduced to the part that says that it must grow by 3.5 percent every year or be declared a recession. What if we hold it steady at whatever trillion-dollar-per-year mark it currently sits?

Ecological side-arguments are often thrown in. For instance, it shouldn't simply be monetarily neutral, it should also be environmentally neutral. Everything, from shirts to washing machines, should be fully recyclable. When you're done with your goods they go back to the factory to be reprocessed into the next round of goods. Packaging should be so degradable that you can simply throw it into the backyard where it will become plant food with no extra work. New raw materials must be reduced at least to the point where there is an equal flow between new material and material heading back to a raw state. We must plant forests at the same rate that we cut them down. We must refill mines and we must mine garbage heaps for raw materials. We must actually, actively, clean the air in equal or greater proportion to the amount we dirty it.

One way this vision works is to increase the number of service-sector jobs and decrease the number of manufacturing jobs. A Plenitude of service. Service, in this way of thinking about it, is virtual stuff. A person having a personal butler might be considered as having a higher status than one buying another car. Indeed, having a limo pick one up would be better than driving. Eating out every night at a different restaurant would be considered better than having a full kitchen with its infinite number of accoutrements. This is not, as you can see, a puritanical, anti-Plenitude vision of the future.

A related version of this zero-growth Plenitude future (which I find personally distasteful) is one that lots of kids like. That is, make all reality virtual reality, goggles and gloves and all. All experience becomes a computer experience. Everything is just data. Once you have the computer and the computer infrastructure, then it is all just cleverness and art. You can start by eliminating all books and magazines. Those are just ASCII and JPEGs anyway. Why go on vacation? Have cameras all around the world so you can see from your desktop the Taj Mahal or a Brazilian beach. Add a chat room and, hey, it might actually be better. Clothing is reduced to avatar clothing. You buy and sell avatars and their accoutrements. And you furnish your virtual house with lavish virtual furniture in whatever virtual neighborhood you want.

This is not that crazy; not only does my eleven-year-old son spend an enormous amount of time furnishing his Sim house now, but he sells his Sim houses for Sim money on the internet. Why have a real Porsche when you can have seven virtual Porsches to drive in any virtual city in any virtual world? And crash them without getting hurt—hit a button and it's back whole again.

This solution works, to the degree it does, because, as I've noted before, we have tiny brains and the experience of driving a Porsche on the web turns out to be only somewhat different than driving a real Porsche on a real street. It's not the same, or as computer gurus always say: Not the same YET. But perhaps that small difference is worthwhile if it saves the world from destruction. Let's take full advantage of our tiny brains. Until the genetic engineers get at them. Perhaps the solution is to make our brains even smaller.

Five: Just make the good stuff.

Five. Let's just make the good stuff—the important stuff. Just the food, the medicine, the housing, the necessary clothing, perhaps the communication system, maybe throw in a transportation system or two, and of course the art. Art being good. And music. Well, maybe not all music.

Take the good parts of the culture, but leave the rest. Leave the bad TV cartoons. The lousy shirts. The put-it-together-yourself particle board furniture. The pulp fiction. Just have handmade furniture, medicines that cure cancer, really good plays and books.

Without even thinking about the legal system this would require, there is a question about whether this works. Doesn't it actually take the full throttle of innovation at every level, including that of producing new cheap shirts and hamburgers, to get the highly rarefied and valuable stuff like new medicines and new communication systems? Is it even possible to have rapid innovation in only designated zones? Isn't innovation linked at the deepest levels? The goods at K-Mart and new antiviral drugs are perhaps are not as separated as they might first appear. It is common, for instance, for the same designers to design both toys and medical equipment. The same brain, operating in about the same way, comes up with visions of both while showering in the morning.

Six: The real problem is too many people.

Six. I like this one, though it is hard to see how we get from here to there without going through some, as they would say on the streets, nasty shit. This starts with the thought that, overall, the Plenitude is fun, exciting, charming, fulfilling, interesting, engaging, enjoyable, sexy, diverse, life-affirming, whatever. What's not to like? Except for this little problem that it would clearly destroy the earth if all six billion people on the earth lived at the levels of plenitude those in the United States currently live at. Even if they lived at the levels of the people of the United States who are in what we call poverty. Even with half the world living at less than $2.00 a day the Plenitude still might destroy the earth.

But what if there were only, say, 500 million people living on the planet? While it is hard to run the numbers, I suspect that this would allow each of us to live in the lap of luxury and still be a small enough number for the world to recover and even regenerate. It could become a moveable feast or an Alice's tea party. The 500 million could move from continent to continent every ten years, allowing the other continents to lie fallow and regenerate for sixty years. Or we could all spread out and each live on our own 1000 acres, content and happy, even as Nature, that curious non-entity, recovers her- or himself.

Will this work? Or do half of these 500 million also have to live at less than $2.00 per day? Or, without draconian measures, do the 500 million quickly become five billion again? Or possibly, do the 500 million discover a new level of heaven on earth that even uses up even more resources? And exactly who is it that would make up the 500 million and what do we do with the other 5.5 billion? Do you picture a future world as diverse as this one, or a homogenous place, all of one tribe, where nothing is wasted on intertribal warfare?

We may get to try some version of this solution in our lifetimes if some of the problems of the Plenitude beat the solutions to the finish line.

```
        Seven: Just love it. We are invisible
          at the scale of the universe.
```

Seven. Lastly, I'd like to present this "solution," which comes up enough for me to realize that a transcendentalism still runs like a deep underground river in our culture.

The planet is 4.5 billion years old. We have been living on it for such a short amount of time, that on a timeline of the earth you can't even draw us with the sharpest .01 pen. And we are part of a galaxy with billions of other stars no doubt with billions of other planets, no doubt with billions of other life forms, no doubt some with their own Plenitudes. And this is just one of billions of galaxies. And the universe is billions of years old and will continue on for trillions of years. And this might be only one of a near infinite number of universes. And they might each cycle, breathing in and out, forever.

At the scale of the universe, both in time and space, we are infinitesimal. We are invisible. It just doesn't matter. And so why not live it in the Plenitude?

But let me end this book with a moral the way good TV shows do. I include this moral because most of the people reading this are probably the creators of the Plenitude. It's us. There is no one else. So here's the moral. It's by a researcher at PARC...

We are the creators of the Plenitude. It's not somebody else. It's us. Yes, we are only part of it, but we are a rather important part of it. We create the new stuff. So here's a little moral fragment that I find useful to carry around with me and insert at the end of most of my talks and books. It is by a researcher at Xerox PARC named Stu Card.

"We should be careful to make the world we actually want to live in." - *Stu Card.*

I think that is good advice.